the joy of
birding

A distinguished bluethroat poses
for the camera.

the joy of
birding

A BEGINNER'S GUIDE

Kate Rowinski

Skyhorse Publishing

Skyhorse Publishing books may be purchased in bulk at special discounts for sales promotion, corporate gifts, fund-raising, or educational purposes. Special editions can also be created to specifications. For details, contact the Special Sales Department, Skyhorse Publishing, 307 West 36th Street, 11th Floor, New York, NY 10018 or info@skyhorsepublishing.com.

Skyhorse® and Skyhorse Publishing® are registered trademarks of Skyhorse Publishing, Inc.®, a Delaware corporation.

Visit our website at www.skyhorsepublishing.com.

10 9 8 7 6 5 4 3 2

Library of Congress Cataloging-in-Publication Data is available on file.

Print ISBN: 978-1-5107-6390-6
Ebook ISBN: 978-1-62636-659-6

Printed in China

This
book is
dedicated to
my children and
grandchild.
Seeing the joys of wilderness
and nature first come alive
for them at the backyard
feeder is one of my most
cherished gifts.

CONTENTS

> **"** What is joy?
> It is a bird
> that we all want to catch.
> It is the same bird
> that we all love to see flying. **"**
> **—SRI CHINMOY**

Introduction

Birder or Bird-Watcher?

DID YOU buy this book to become a birder or a bird-watcher? Maybe you don't know yet. Maybe you didn't even know there was a difference. You just know that you feel a little rush of excitement when you see the first robin in the spring, or that you thrill at the sound of the cardinal singing in the nearby patch of woods. It doesn't really matter anyway. The joy of birds in our lives takes many forms.

My first memory of birds was as a very young child. My mother kept a small feeder outside our kitchen window in northern Wisconsin. Mom was from the South and had not taken well to the cold Wisconsin winters. From January through March, she did not go outside unless it was absolutely necessary. But every morning, I watched as she ventured out into the cold brightening dawn, a heavy coat and boots thrown over her nightgown and slippers. She went out to the porch and filled a scoop with a rich blend of seeds and nuts. Then she carried it around the house to the kitchen window. My dad usually kept a little path shoveled for her, but often the night's snows had filled it in. Mom stepped through, regardless of the depth, to an old wooden shelf feeder that my dad had made in his shop.

The feeder had a little roof for keeping snow off the seeds, but my mother always took off her gloves briefly to remove stray snow that had accumulated in edges and corners. By the time she got back into the house, stomping snow off her boots and pulling off gloves, the scattered birds had already made their way back

LEFT: This black-capped chickadee enjoys a snack.

to the feeder and were busily munching on their breakfast. I was stationed at my regular spot, sitting inside the kitchen sink, to watch their morning ritual. My mother would come and look over my shoulder, nodding with satisfaction. "Poor little things," she would say. "If they have to survive all winter in this weather, the least we can do is get them a good meal." My mom was not an avid birder—she didn't know why some birds headed to friendlier climes when it turned cold while others did not. She just saw kindred spirits in those little birds—creatures doing their best to get through a long winter.

Mom's primary customers were the black-capped chickadees. They generally came to feed as a small group. One would arrive on the scene—I always imagined him to be the scout—quickly to be followed by others. Soon a full-fledged chickadee party had broken out. I loved their busy, social back-and-forth from pine branch to feeder, hopping away to make room for others, back again for their turn. They were in constant motion, performing a graceful ballet of movement, often hanging upside down or grasping a branch sideways in a perfect display of bird gymnastics. Of course, there were other little birds whose names we didn't know. I watched the birds for hours at a time, noting their habits and their squabbles, which species were a nuisance and which were shy. I came to love those little birds, imagining that I knew each of them personally. I had become a bird-watcher.

In high school, I was introduced to birds in a more scientific way. My freshman biology teacher, Mr. White, was a passionate birder. He kept huge charts of bird species in the classroom and required us to memorize every order, family, tribe, and genus. We learned to tell the difference between the Passeriformes and the Apodiformes. We could identify the wing shape of the Falconiformes and the distinctive posture of the Piciformes. Bird recordings played in the background of every class. Tests included identifying every bird by silhouette and voice. By May of that year, when I noted the arrival of our robins before I actually saw them by their familiar *cheerily-cheer-up* and could identify the distinctive *chippee-chippee-chippee* announcing the arrival of the warblers, I knew more about birds than I thought it was reasonable for any fourteen-year-old to be expected to know.

In college, my focus turned to birds in the field. Though I was an English major myself, I often found myself hiking through the woods with my boyfriend, a biology major who was conducting fieldwork on wilderness forests. I thought I had forgotten the lessons that Mr. White had drilled into our heads, but soon found that I was able to place most birds into their proper family by their posture, wing shape, or voice. While Jim was off counting trees or collecting water samples from the rivers, I began making a list of the birds I saw. I started small, just noting in my journal when I caught sight of an unusual hawk or spotted a magnificent great horned owl. But soon I found myself noting *all* the birds I saw. I stopped over to my mom's house to pick up my old *Peterson's Field Guide* so I could identify birds more accurately. Pausing to watch the familiar backyard birds at Mom's feeder, I realized that I could add all the birds *I had ever seen* to my list. Little did I know I was following the time-honored obsession of people who noticed birds—to collect a list of every bird I had ever seen.

I had become a birder.

As my family grew, my life list also grew, expanding to include species from several different states and countries. The truly *new* bird was becoming rare, and I lost interest in the passionate focus on adding the five-hundredth bird to the list.

Introducing birds to my children brought the passion back for me. Through their eyes and the squeals of delight at the antics of chickadees or the elegant flight of hummingbirds, I regained the joy I'd found in simply being present in the lives of these little creatures. I slowed down again to really stop and enjoy the birds. I found myself bringing my coffee to the window near the closest feeder so I could watch them over my breakfast. I placed feeders where I could sit quietly and study their habits. After a lifetime of birding, my transition was complete.

I was once again a bird-watcher.

Whatever your reason for learning about birds—for companionship on a cold winter day or for the joy of spotting a rare warbler on a lone hike at your local park—take the time to learn the basics about their habits, their lifestyles, their likes and dislikes. Knowing about their lives will increase your enjoyment of these delicate little creatures.

This book is designed to provide you with the three fundamentals you need to become a birder: First, learn to understand the world from a bird's point of view. Knowing their likes and dislikes, reasons for choosing one area over another, and feeding and migratory habits will put you in their "shoes" and give you a new way of looking at the world. Then, when you see a bird behaving a certain way, even if you don't know what it is, you can make an educated guess about what he is doing and why.

Second, learn the basics of identification. If you have a rudimentary understanding of the categories and families that birds belong to, and the characteristics of each, it will make it a lot easier to place them. Being able to say *That's some kind of thrush* puts you way ahead when it comes to trying to pin down a particular species.

Third, really get to know the species in your own neighborhood. Learn about the habits of the local birds in your area—it will go a long way toward helping you understand the population as a whole. And there is something remarkably satisfying about knowing that the tiny little bird on your ledge would really prefer it if you would give him hulled seeds, or that no matter how badly you want to attract that bluebird to your feeder, his dinner plans are elsewhere. Or that lovely little hummingbird has a silken nest somewhere up in the trees that she is going home to after she leaves your flower bed.

Knowing how to provide a good habitat for your birds, what to feed them, and how to provide for their other needs will bring with it a wealth of satisfaction and enrich your world. There is nothing like taking a few minutes to watch your birds going about their business in your backyard to put the whole world back into perspective.

Enjoy!

A Note

I AM a bird hobbyist. I am not a biologist or a researcher. I just like birds. I enjoy hosting them in my yard, and I like watching them overhead. I get a kick out of knowing one warbler from another, and that come spring, yet another wren will nest in a planter on my porch. I am thrilled when I see nesting geese on our lake, and can't get enough of the nuthatches and woodpeckers at the suet feeder. Like the millions of other bird lovers around the country, I am concerned with providing a healthy environment for birds. The loss of land to development and the other human habits that endanger birds drive me to do what I can in my little corner of the world to provide welcoming habitat.

I offer my thanks to the wonderful people in the Charlottesville birding community. Many people were helpful in answering my questions about certain species and tricks of the trade. I would also like to thank my brother Cary for his generosity in sharing his interest in Florida birding with us, as well as the photo contribution he made to this book.

If there are mistakes in these pages, they are mine and mine alone.

LEFT: Canada geese

It's lunchtime for this robin family.

the joy of
birding

> **"** My head is full of birds. **"**
> **—JEN ROWINSKI, age five**

① Understanding the Bird

WHETHER YOU want to become an expert at bird identification or just get to know your neighbors a little better, your first step should be to learn a few basics about where birds live and the reasons behind their choices. Identifying a bird begins with understanding the ecosystem around you. Why is this bird in this place? Is it here to eat or to build a nest? Or is it just passing through? What is it saying? Answering these questions reveals a world like no other—an alternative universe of high-drama courtships, joyous births, bittersweet departures, and heroic battles for life—all taking place in the treetops and shrubbery of your own backyard.

Where Birds Live

Every species of bird makes very specific choices about where they like to spend their time, what they prefer to eat, and who they prefer to share their territory with. Even the wide varieties of birds in your own backyard are sharing that space because of predetermined conditions present in that habitat. To understand the birds in your own backyard, you need to be in

LEFT: A song sparrow

tune with the seasons, and aware of how every change in food availability and cover impacts the distribution of birds in your area.

The Range

One cold winter morning, my ten-year-old son came into the kitchen and said, "Hey, Mom, there's an oak titmouse at our feeder."

"No, there isn't," I said.

"Yes, yes," he replied, pushing the field guide into my hands. "See?" He pointed at the picture of the little gray bird. "Our titmice have brownish bellies. This one doesn't. He's all gray."

I followed him to the feeder and watched as he pointed out the bird he was observing. It was mostly gray and white, a slight variation of our normal rusty-brown belly. "Good eyes!" I said, "but no, that's just one of our tufteds. Maybe it's a youngster, or maybe it just washes its belly."

"Mom!"

"Take another look at your guide. What is the oak titmouse's range?"

He looked again. The oak titmouse was off somewhere in sunny Baja California, no doubt enjoying its breakfast in someone else's backyard.

Lesson One: *Understand where birds live.*

With over 10,000 species in the world and nearly 1,000 here in the United States, you'd think there would be a lot of jostling for elbow room. But birds have an elegant system for sharing all that space. Unlike humans and other mammals, most birds don't call just one place home; instead, they have ranges. The range of a species is where it is regularly found. Species have summer or breeding ranges, winter or nonbreeding ranges, and migration areas where they might be spotted while on the move. Some have year-round ranges. That's why the range maps in your field guide are so important. They help you to determine whether the bird you are looking at could reasonably be expected to be there.

As you become familiar with your own area, you will learn to identify how birds fit into your region. You may find yourself asking a lot of questions. Do those little blue birds live here, or are they just passing through? I had lots of robins; where did they go? We had two weeks of amazing warblers; what happened to them? As you gain knowledge and confidence, you may even be able to spot the rare sighting of one that has actually gone outside of its range.

Range maps describe each range by the habits of the birds in those areas:

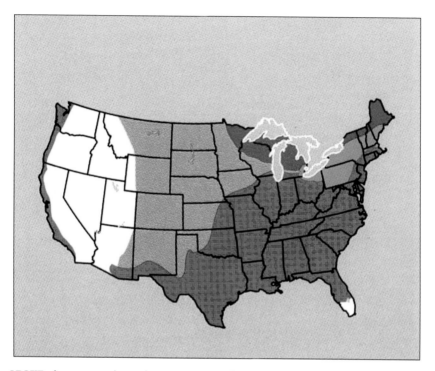

ABOVE: A range map shows where a species spends its time.

Year-round residents: These are your regulars. They set up shop in the general vicinity and can reasonably be expected to be seen in your area for most of the year. That doesn't mean they stay in your backyard, though; they may roam throughout the region as food supplies and nesting requirements change. Chickadees may enjoy the summer in your woodlot, but move closer to the feeder for the winter. Juncos leave the mountain and "come to town" for the winter. Robins may still be around too, but cold conditions have reduced the appeal of your yard, with those yummy earthworms no longer abundant.

Summer residents: Every species has requirements that make a region the ideal place to raise a family. Summer residents generally come to establish their nest, stay as late as October to take advantage of the abundant food, and then head off to more-southern regions for their nonbreeding season.

Winter visitors: You might think that Florida is the place to go for a winter visit, but many birds think even a relatively small change in their environment is enough. Many of our winter visitors have summered in Canada, so the Virginia winter looks pretty inviting to them!

Migrants: These are the birds that we look forward to catching a glimpse of in spring and fall. You might wake up one morning to a tree full of yellow warblers, just passing through to other areas where they will spend their season. Enjoy them while you can! They will feast for a few days and then continue on their trek. Making a note in your calendar of when you see these beauties will help you to be ready and watching for them the following year.

Territories

Within a bird's range is its territory, the area that he has decided to call "his" for the season. Bird territories are usually temporary. Even for birds that return every season, boundaries shift and change as other species move in and out. Males spend a lot of time and energy staking their claim, defending it from intruders, and announcing to others the boundaries of their empire.

Birds often have to adapt to the course of human events: A forest taken down over the winter for a shopping mall may require a bird to look at a local park that has traditionally belonged to someone else. A marsh drained for a subdivision may require a sudden change in plans.

All your own small decisions about your yard or woodlot affect your visitors. We had a big oak in a central location that had died, and we began to grow concerned with its stability and its proximity to our house. Every storm caused a shower of brittle branches. We sadly took it down, understanding the damage we had done to our population of cavity-dwellers. On the other hand, the addition of cherry trees and blackberries bushes to our yard, although not particularly beneficial to our own food stores, became a boon to birds that had overlooked us before.

Layers of Habitat

Have you ever thought about why chickadees and warblers are in the same yard? Or whether those cardinals mind having those jays around? The fact is, while your yard seems like only one environment to you, to the bird there are actually several habitats represented. Every species has specific needs that can be met without interfering in others' food sources or nesting requirements. Nature has created an elegant system that allows you to host a multitude of different types of birds in your own small backyard.

The reason that so many birds can comfortably make a temporary home in your backyard is because every environment has a series of separate and distinct habitats that perfectly suit individual species.

The canopy: Sometimes called the *overstory*, the canopy is the highest vegetative layer in the forest. The canopy is filled with leaves from the forest's large, most mature trees. Extending 60 to 100 feet high, this habitat is rich with beetles, caterpillars, and leafhoppers that make their homes on the highest layer of the treetops, providing a rich environment for foraging. Birds like blue jays, owls, hawks, and eagles live and work in the top layer of the forest environment. Some songbirds like the yellow-throated warbler build nests here. The thick leafy canopy provides protection for the open nests of these birds.

The understory: The understory of the forest is a shady world of younger trees and shade-tolerant varieties. Only 50 percent of the total sunlight can get through the canopy to the understory. Just below this top layer of sun-

LEFT: Your yard represents four distinct layers of habitat. Different bird species eat and live in treetops, the shady understory, in shrubs, or on the ground.

BELOW: Eagles and other birds of prey find treetop perches ideally suited for spotting fish and small mammals below.

gathering leaves, this habitat hosts a completely different environment for bird species. The understory features the trunks of the mature trees that leaf out the canopy, as well as the smaller, less-mature trees of the same species. The rough bark of these trees has split apart and developed cracks. These trunks are home to a staggering variety of worms, beetles, ants, and other insects. That's what makes it the ideal habitat for woodpeckers and flickers.

BELOW: The petite ruby-crowned kinglet gleans small insects from leaves and small branches.

Smaller cavity-dwelling birds such as nuthatches like this environment too. But instead of hammering away at the trunk itself, you can often see them scooting down the trunk headfirst, gathering the insect eggs, beetles, spiders, and small caterpillars that the wood-boring birds missed. Chickadees find a multitude of hiding places for seeds and insects in the crevices of the mature bark.

Varieties of shade-loving trees also make their home here. In my backyard in Virginia, this layer features varieties such as shadbush, sourwood, dogwood, and redbud, which attract a range of seed-loving birds.

The shrub layer: The shrub layer is categorized by leafy plants that extend no more than 6 or 7 feet high. Azaleas, rhododendron, and mountain laurels are common varieties in this layer. This habitat is the ideal feeding ground for many small flying insects. Gnats and blackflies are

RIGHT: A white-breasted nuthatch working down the crevices of a tree trunk in search of insect eggs and seeds.

plentiful here in the spring and provide a tasty, high-protein diet for a variety of birds that are preparing for nesting. The rose-breasted grosbeak lives in the shrub layer, preferring to build its nest in shrubs 5 or more feet off the ground. Robins, chickadees, warblers, quail, sparrows, finches, and cardinals also find wild and cultivated fruits, berries, earth-worms, and insects such as beetle grubs, caterpillars, and grasshoppers in this layer.

The ground layer: This layer features plant varieties that bloom mostly in the spring, along with lichens and mosses. Dead logs also host bark beetles, larvae, carpenter ants, and earthworms, as well as an array of

BELOW: Shade-loving trees like dogwoods and redbud help form the understory layer.

LEFT: The northern flicker is as comfortable with dead wood as its woodpecker cousins, but actually forages more often for food on the ground. You can often find it hammering on the forest floor in search of ants and beetles.

RIGHT: This female northern cardinal is at home in the safety of this tangle of berry bushes.

BELOW LEFT: Dark-eyed juncos are ground foragers, but watch for them perching on low branches or shrubs.

BELOW RIGHT: Tall grasses and leafy undergrowth provide a natural hiding place for this white-throated sparrow.

spiders, centipedes, and slugs. Ground-foraging birds find an abundance of food in this layer.

Other Habitats

Meadows and Grasslands

Some birds, such as the red-winged blackbird and the eastern meadowlark, along with bobwhites and field sparrows, enjoy the open-air market of the meadow. Grasshoppers, insects, and seeds make up their diet. This is also a great place for predators, however, so ground nesters like the bobwhite have to be constantly vigilant. They take the precaution of sleeping in groups and camouflaging their nests with grasses and other nearby vegetation. A great hunting ground for small birds and rodents, owl and hawk sightings are also common here.

Marshes, Wetlands, and Waterways

Some birds live and work only where there is an abundance of water. Marshes and wetlands are important habitats for these species, and a great

ABOVE: A flock of red-winged blackbirds congregate in an open pasture.

place to bird-watch. Marshes generally support large populations of birds that are uniquely suited to waterside living. Because water height varies from year to year, some nesting birds create clever floating nests that allow them to adapt to changes. You can expect to see herons and egrets, which have ample sources of food in the marsh environment. Look for their nests high above the water. Saltwater environments are homes to yet another set.

ABOVE: A northern bobwhite quail blends easily into a background of leaves and dry grasses.

What Birds Eat

Birds live and work where the food is, and they have to work hard to find all the food they need. Their high metabolisms demand fuel, and lots of it. Birds eat anywhere between 5 and 300 percent of their body weight every day. Tiny birds like chickadees and wrens consume at least half their body weight in tiny insects every day. Hummingbirds will eat up to three times their body weight in nectar every day.

Multiple species can live in habitats together because different aspects of the habitat provide for different needs. Likewise, different seasons provide different food sources, and birds have adapted to take advantage of

ABOVE: One of the most common of water birds, the great blue heron makes its home wherever the fishing is good!

ABOVE: Eastern towhees are perfectly at home in thick undergrowth, surrounded by berries and small insects.

those sources. Your yard is appealing specifically because of what is going on there in any given season.

Spring

Most species are on the move in the spring when insect populations are beginning to explode. The abundance of available animal protein makes it possible for mating birds to prepare for nesting season. Many birds are specialists, relying primarily on a particular category of insect. Others are more opportunistic, and will accept a varied diet. Some of the favorite food options include:

Caterpillars: Many birds rely on at least some insect protein, and spring provides it in spades with caterpillars. Nuthatches, warblers, cuckoos, mockingbirds, and blue jays are among the species that help us by munching on

RIGHT: Dragonflies are great insect predators in their own right, but are also a juicy prey for birds.

tent caterpillars, but cardinals, grosbeaks, tanagers, and many others also love a meal of caterpillars.

Mosquitoes and other flying insects: Some birds specialize in catching insects on the fly. Martins have long had a reputation for keeping mosquito populations under control, but several other varieties of birds also do their part. Swallows, flycatchers, phoebes, warblers, and waxwings are among the many birds that like mosquitoes.

Beetles, grubs, and spiders: Ground foragers specialize in these bugs. Blackbirds, bluebirds, thrushes, wrens, starlings, and towhees are among those that prefer this category.

Earthworms: Many ground foragers hunt earthworms, with the American robin and other thrushes among the best-known customers.

Water insects and larvae: Certain birds specialize in waterborne insects. Some birds hunt the air over the water for these hatches, while others will go into the water for their food. Kingfishers, crows, robins, flycatchers, ducks, and shorebirds are among the many birds taking advantage of bodies of water for their food source.

Bark insects: Woodpeckers, chickadees, nuthatches, wrens, and sapsuckers spend much of their time examining and picking at the crevices of tree trunks and branches for the insects that live there.

BELOW: Downy woodpeckers are small and acrobatic enough to take advantage of larvae clinging to thin branches and weed stems.

Summer

Many birds that have spent the spring eating insects and other animal matter in preparation for nesting are now looking to the fruits and seeds of summer. Although many songbirds will continue to collect insects to feed their growing youngsters, they may begin to seek out berries and other fruits for themselves. Others, such as the cedar waxwing, prefer a diet that is almost entirely fruit. In summer, watch for birds looking for the following:

ABOVE: The American robin spends most of the spring and summer in search of succulent earthworms, supplementing its diet with fruit and berries.

BELOW: If mayfly larvae are lucky enough to hatch from the water, they must dry their wings before flying off, making them an easy catch for birds.

Beetles: Sparrows, bluebirds, orioles, and downy woodpeckers are just some of the birds that are happy to munch on beetles. Cardinals, grackles, starlings, and robins are all helpful in controlling Japanese beetles.

Cicadas: When summer is filled with the sound of cicadas, robins, grackles, and buntings will take full advantage of this plentiful food source.

Flying insects: Specialists in flying insects are now enjoying the peak of their food supply.

Shrub fruit: Blackberries, raspberries, blueberries, grapes, and strawberries will attract mockingbirds, orioles, and cedar waxwings, among many others.

Tree fruit: Cherries, Juneberries, and mulberries are also attractive to fruit eaters.

Autumn

Autumn brings its own set of treats just in time for birds that are preparing to transition back to cold weather. The harvest is on, and birds are ready to take advantage of the changing season. Hardy fall bushes and shrubs are offering up native fruits and nuts with higher fat content.

BELOW AND RIGHT: An abundance of summer fruit is sure to attract a variety of songbirds.

Berry-bearing autumn shrubs: Winterberry, spicebush, sumac, and Virginia creeper attract robins, bobwhites, kingbirds, catbirds, and fly-catchers.

Decorative trees: Dogwoods and Bradford pears produce plentiful fruit and make a great meal for bluebirds and grosbeaks, among others.

Wildflowers: Weed seeds are abundant now, and sparrows and finches make the most of them.

Late-blooming flowers: Salvia, hollyhocks, and lobelias are greatly appreciated by hummingbirds as they start their fall migration.

Ground insects: Fallen leaves create great ground cover for hiding a wealth of insects, grubs, and other tasty treats for your ground foragers.

Winter

Depending on where you live, winter brings a variety of challenges. For northern areas, snow covers much of the ground-foraging possibilities, and leftover fruits, berries, and tree seeds have been largely picked over.

Late-blooming flowers and fall berries are a boon to migrating birds.

RIGHT: The tufted titmouse is a master hoarder, usually shelling a seed before hiding it away for later use.

As you head south, where the weather is cold but snow cover is minimal, birds will still find opportunities in the conifers and tulip trees.

Now is the time when the ground cover and the shrub and tree assortment in your backyard may distinguish it in the eyes of the birds that are planning to stay for the winter. Conifer trees, standing clumps of ornamental grasses, and evergreen shrubs like boxwood and holly are inviting places for birds this time of year. In addition to providing shelter from the wind and snow, these areas also make great hiding places for birds looking to escape predators that are also on the hunt for food.

Leftover seeds: Trees, garden flowers, and weeds are thoroughly inspected for these leftovers.

Insects: Even in the dead of winter, insects are hiding in tree crevices, beneath dead leaves or loose bark.

Winter berries and fruits: Holly berries, dogwood, rose hips, and bittersweet vines that retain their berries throughout the winter are a boon to birds this time of year.

Conifer seeds: Chickadees, pine siskins, and nuthatches are among the birds that may seek the small seeds in pinecones.

ABOVE: Winter holly, dogwood berries, conifer seeds, and other nuts make up much of birds' winter diet.

Birds and Water

Birds need water, and they are especially adept at finding the most unlikely sources for it. Dewdrops and rainwater caught on leaves are often an adequate source for small birds like kinglets and hummingbirds, while puddles and small running brooks are a great source for others. Most birds rely on the immediate supplies of water in their neighborhood. Hot weather and drought conditions can wreak havoc, however, so birds are always on the lookout for good sources of water.

ABOVE: Even a droplet of rain on a flower petal can be a good source of water for hummingbirds or kinglets.

BELOW: It doesn't take more than an inch or two of water to provide a satisfying bath for this American robin.

How Birds Communicate

Dawn is my favorite time to listen to birds. The first thing I do when I wake up on a fresh spring morning is to listen to the sounds outside my window. Keeping my eyes shut, I wait for them to begin. Chirps and whistles and melodies come from the tall tulip trees in the backyard and I can hear a familiar *fee-bee-fee-bee* from over by the dogwoods. They are pretty sounds and, although meaningless to most of us, they can tell us a great deal about our world.

Over the years, I have started to listen to the sounds of my backyard with more purpose. Who is already out and about? How many birds are at the feeder? Are there any squabbles going on yet? Has someone seen a cat or sent up an alarm about a hawk? Do I hear any new voices? Is there a new bird in the forest preparing to set up housekeeping? Are they commenting on the weather?

I work in an urban area. One evening, I left my office especially late, my head full of worries and concerns over "office stuff." As I reached my car, I heard a sound in the vicinity of a tree near my parking garage. A beautiful heartfelt song was coming from the branches. I walked over and listened for a few moments. No need to see him—I knew who he was. A lovesick male mockingbird, still unmated a little late in the season. In the moonlight, in a pine tree surrounded by highways and skyscrapers and steel, he was pouring out a beautiful song, hoping against hope that somewhere in the concrete jungle a female was waiting for him. For a few moments we shared the evening together, two beings with a common understanding. When I got into my car, my office worries had been replaced. My head, as my daughter used to say, was full of birds.

BELOW: Early morning is a magical time for bird-watching.

ABOVE: The male American redstart is a small warbler that can be recognized by its simple *chewy-chewy-chewy, chew-chew-chew.*

The variety in bird vocalizations is remarkable. It is a complex and astoundingly beautiful form of communication. Bird sounds take many forms and serve many functions, including attracting mates, warning away others, communicating with their own species, and announcing the boundaries of their territory. It may seem impossible to try to decipher the seemingly endless variety of sounds, but there are a few basic principles that can help you to understand what birds are saying.

The Song

Songbirds make up over half of the world's bird population. Birdsongs are a series of notes, both short and long, that have a pleasing musical sound. These are the sounds of spring and summer. The male songbird uses a song to attract a mate. The mating song tends to be complex and elaborate, often with many variations. The nightingale, it is said, has over three hundred different love songs at its disposal. Still others, like the cowbird, have been recorded using over forty different notes. In most species, females listen, but do not respond to these songs.

Males also use their songs to tell others that this is their territory. These songs tend to be shorter and simpler. They are meant to tell rival birds where they are and what the boundaries of their property are. There are gaps in these songs, when the male pauses to listen for answers from possible rivals.

ABOVE: The house wren has a number of calls that it uses to raise an alarm. Listen for it chattering and scolding as you approach its nest.

Although songs are used mostly by males, some species, like the northern cardinal, use songs to communicate with each other throughout the season. Both the male and the female sing. Some birds have multiple versions of their song, while others repeat certain phrases over and over. Still others have single-note songs.

Dawn is a busy time for birdsong. Maybe it is because the air is relatively still and the sounds carry better than later in the day. Songs sung at dawn tend to be clearer and more elaborate than songs performed later in the day. There are also songs sung *sotto voce*, whispering songs that are often soft renditions of the bird's regular song, sung during migration, on the nest, or in inclement weather.

The Call

Calls are short vocalizations that are used by all birds, both male and female. They are used year-round and have a very specific purpose.

Contact calls are the most common, used to make contact either with a mate, with other members of their species, or with the mixed flock that they are migrating with.

Flight calls are used by some birds during night migrations, or even when they are just flying from perch to perch.

Alarm calls warn others about the presence of a predator. Some alarm calls are just louder, more-urgent versions of the standard contact call, while others are high-pitched and repetitive sirens.

Begging calls are whiny cries performed by juveniles still looking for a handout from their parents.

Some birds can be distinguished from their similar counterparts by their call. We have a mixed flock of black-capped chickadees and Carolina chickadees. Their appearance is virtually identical, but we can identify them by their different call notes. The Carolina chickadee has a higher pitch and a faster pattern than the black-capped, and their songs are also different.

On the other hand, our finches are a hopeless blur to us. If we don't actually see them, we usually have a hard time figuring out who is in the trees. Is it the house finch or the purple finch? Or could it be a warbling vireo?

Mimicry

Some birds have a remarkable ability to simulate other bird sounds, and even non-bird sounds. The catbird mimics a number of different birds, giving a rival the impression that the territory is already full. The blue jay does a great imitation of the red-shouldered hawk, a skill designed to send its rivals at the bird feeder running for cover. Starlings, mockingbirds, and goldfinches are also excellent mimics.

Non-vocal Communications

Birds are not limited by their vocal skills. Some birds, like the downy woodpecker, drum with their beaks to attract females. Clicking noises made by snapping bills and thrumming sounds made by loud wing clapping are also

BELOW: Is there a tree full of birds in the backyard—or is it just one rascally mockingbird?

used by males to elicit attention. The ruffed grouse beats the air with its wings, and the American woodcock's wings make a twittering sound when it flies.

How to Make Sense of What You Hear

I have an understanding of the general tone of the birds in my yard. I can take their emotional temperature by understanding the tension in their calls, the insistence of their songs, and the tone in their *chips*. I don't know who everyone is, but I can gauge how things are going by the sounds that I hear.

Simple *chip* notes can be learned by paying attention at your feeder; get accustomed to the *chip* notes of your familiar birds. Watch your dark-eyed juncos as they hop around under your feeder. They *chip* away mindlessly as they feed, reminding me of a great-aunt that never stopped talking to herself. See which birds seem to call an alarm; did that chickadee just tell everyone to run for cover? Listening to the little noises is a great way to get to know your birds better, and it will help you to more quickly notice when a new *chip* is present, giving you your first hint that a new species is visiting.

More-serious identifications take some practice. Ask a birder how he identifies bird sounds. The answers can be quite comical, especially for the untrained ear. Experienced birders refer to the tempo, the pitch, the cadence, and the quality.

But that's just the beginning. Birders also create phrases to describe birdsongs. They suggest you listen for a wren's *do-it-again, do-it-again* phrasing, or the *cheer-cheer-up* of the robin. I often cannot hear what others are describing until I let their words fall away and try to pick up on the pattern of the phrasing. Then my ear finally picks up the cadence. Invariably, however, I apply my own words to the bird's song. While I can agree that the tufted titmouse says *peter peter peter,* the way I tell the difference between it and the Carolina wren is that the wren says *cheeseburger, cheeseburger.* Don't ask me why; that's just how I remember it. Of course, some birds are named after the sounds they make. Who could mistake the phoebe, which sings *fee-bee, fee-bee* over and over?

Take a stab at describing the quality of the birdsong you hear. At first, it may seem artificial and unfamiliar to say that a birdcall is metallic or

a song is squeaky. It's like the first time you try to describe the taste of a particular wine. But you may notice some common themes that resonate with you. Thrushes have fluid notes that sound fluty. Trills can be described as loose or rapid and dry. Some birds have a buzzing *zzzz* sound. You can also identify tinny chirps, rasps, and nasally whistles. Once you are able to put names to the different sounds you hear, the short metallic sound in the *chip* of a northern cardinal will sound very different from the longer, high-pitched *see* of a nervous chickadee, and you will be able to say exactly why.

There are a wide range of recordings available to help you practice your birdcall identification skills. I like the ones that pair similar-sounding birds together. It gives me a frame of reference for the type of bird that makes the sound, and allows my ear to pick out the differences. Practice is really necessary to become really good at identifying birds by their sounds. Start with the birds that you know. Learn the robin's repertoire, and then add a cardinal's song and a chickadee's sounds. With practice, you will be able to identify several birds in your own yard.

Then venture farther afield. The very best way to learn is to go out into the field with an experienced birder who can tell you what to listen for. Many experienced birders speak "bird" and will be able to readily identify a wide variety of sounds. When they hear a song, they follow it to its location and try to match it to the singer. With a bird in sight, they will wait for it to call again to confirm a match. This kind of search and discovery tends to stick with you better than any other kind of mnemonics or memorization.

Where Birds Nest

Nesting season is the most important time of the year in the bird world. Males search out mates, and, depending on the species, males and females may search out the best location for the nest together. What is considered a perfect building site varies dramatically between species. A number of birds, from large woodpeckers to miniature nuthatches, use holes in trees. Others want the nice crook of a small tree for their nest, while still others look for a penthouse location, as high in the tree as they can get. Some birds will reuse their nest for another brood, while others refuse to use the nest more than once. Still others are perfectly happy to "borrow" someone else's nest.

Nests are feats of remarkable engineering, and nesting materials include a wide array of materials. Twigs, grasses, and roots or vines are common building blocks. Some birds, like the petite hummingbird, weave intricate nests using thistledown, spider silk, and moss. Others, like the robin, create a multilayered nest of twigs and grasses with a binding coat of mud and a lining of soft, dry grass.

Although there are more varieties of nest-building techniques than you can imagine, most North American nests can be categorized roughly into these types:

Ground Nests

The simplest of nests, this is a shallow indentation scraped in the ground, often used by shorebirds or pheasant, grouse, and partridge. Most birds will scrape or stomp down an area with their feet, although terns will find a sandy location and rock their bodies back and forth to create the shallow depression. Some birds will line the area with grasses or dead leaves, while

BELOW: The eastern meadowlark's ground nest is woven right into the landscape and camouflaged from view with an arched opening.

a number of ducks will actually "feather their nest" with soft down plucked from their own bodies. Because ground nests can be quite exposed to predators, nature put in a few safeguards. Both the eggs and the nesting female tend to take on a camouflage color that helps them blend into the surroundings. Babies are born fairly mature, allowing them to leave the nest quickly, as early as twenty-four hours after hatching.

Cavity Nests

Cavity-dwellers include both birds that are capable of excavating their own holes in trees, and those who rely on existing holes to make their homes. Woodpeckers, which dig their own cavities, often line the bottom of their

BELOW: A nest filled with baby swallows, awaiting their next meal.

holes with wood chips before laying their eggs. Other cavity-dwellers, such as eastern bluebirds, which rely on finding a vacant hole, may line their holes with wood chips, leaves, moss, or animal hair. Cavity nests have the advantage of being warm, dry, and sheltering. Unfortunately, they are fairly easy for predators to access, so cavity-dwellers have some ingenious deterrents. Cavity-dwelling birds may smear pitch or resin around the entrance to deter visitors.

ABOVE: This wood stork's nest may be used by the family for several years.

Platform Nests

The most recognizable version of this nest belongs to one of many large birds that raise their young high in the trees. These complex nests, also known as *aeries*, are built by eagles and ospreys of branches and twigs. Often placed in the highest tree in the area, these nests are built to last, as bird families use the same nest year after year. Other birds use the platform style too. Common loons and some waterfowl build platform nests on the ground, and still other waterfowl actually create floating platform nests.

Cup Nests

When most people think of a bird's nest, this is the style they think of. You can find cup nests all around your yard—perched in the crook of branches, tucked into one of your hanging planters, perched in a gutter, or in the eaves of your garage—any sheltered place is a

RIGHT: A perfect clutch of American robin's eggs; a female may produce as many as three clutches per year.

possible building site. Built by most songbirds, these structures are made in a series of layers. The female starts with the outer walls. She forms this of grasses, leaves, and twigs, often glued together with mud. Then she lines the inner layer with soft materials such as grass, feathers, or moss.

Different species have variations on this basic structure. The female hummingbird's delicate nest favors spider silk as her "glue," and is a miniature work of art. Crows, which follow the same basic blueprint, create large, messy versions. Swallows build cuplike nests out of mud, often attaching them to the sides of houses or inside chimneys.

ABOVE: A buff-bellied hummingbird sits in her nest of bark and grass fibers, suspended delicately from a branch.

Pendant Nests

One of the least-common nests here in North America is the pendant, tightly woven of grass and other flexible plant materials. It is a sack-like structure and hangs precariously off a small branch. Favored by Baltimore orioles, it is surprisingly safe from predators. It has a small opening at the top or side to allow access by the parent, but it doesn't allow easy entry by a typical nest marauder. On the other hand, hanging from this little branch has definite disadvantages, as strong breezes give this little nest a wild ride.

How to Identify a Nest

If you want to know who lives in a nest, it is probably going to be a matter of watch and wait! First, take note of where the nest is. This will give you an idea of what you are looking for. If the nest is in the cavity of a tree or in a nesting box, you have narrowed down the possibilities to typical cavity-dwellers. If it's in a tree, note the height, size, and shape of the nest. You may find eggs which can help you identify what type of bird is nesting there. But the best way to know for sure is to see the parents coming and going.

If you have noticed a nest in your yard and are curious about what lives there, be respectful of the inhabitants. Nesting is a nervous time for the birds, so don't do anything that will disrupt the process. Don't peek into active nests first thing in the morning or after dusk; that's when the mother is most likely to be home, and you don't want to scare her off. Carolina wrens regularly use my hanging planters for their nests. I am always amazed by how quickly these nests seem to appear, so I always take a peek before I start watering.

Baby Birds

Unless you are monitoring the nests in your yard, it is really best to keep your distance from any active nest. Most songbirds incubate their eggs in

ABOVE: As fragile as songbird nestlings may seem, they will be ready to face the world on their own within twelve to fourteen days.

just around two weeks and most fledglings are ready to go in another ten to twelve days. So even if you find a nest in an inconvenient location, try to be patient and stay away until the babies have safely fledged.

Weather

"Birds flying high in the sky predict fair weather."
"Birds on a telephone wire predict the coming of rain."
"Seagulls stop flying and take refuge if a storm is coming."
"If birds come to feed during a storm, it will rain for a long time."
"If they perch in the trees to wait it out, it will clear soon."
"If a rooster crows at night, there will be rain by morning."
"When you hear the 'rain crow,' you know rain is just around the corner."

Birds have long been considered harbingers of rain and storms; humans look to their behaviors for reliable weather predictors. In fact, like most folklore, there is a touch of truth in some of these common sayings. Low-pressure systems and temperature changes that forecast weather probably do create an observable change in behavior.

On the other hand, the "rain crow"—otherwise known as the yellow-billed cuckoo—probably doesn't know any more about the forecast than you do. Yet somehow his lonesome call is often associated with that silent time just before a summer shower. It is a favorite sound of mine—a haunting song from my shy woodland neighbor.

Rain

"Fine weather for ducks" is a common way to discuss a rainy day. And why not? Birds, as you have probably noticed, are waterproof. Their outer feathers protect the softer insulating down feathers underneath, and they can shake off a spring shower quite easily. During a storm, birds will often settle under the shelter of shrubs or leafy branches. But after a long dry spell or in hot weather, we often see birds perching out in the rain when they could easily take shelter. After all, what's not to like about a refreshing shower on a warm day?

Beautiful weather with high pressure and warm conditions provides great flying conditions for migrating birds. But if bad weather is in the forecast during migration, it's a good idea to keep a "weather eye" on the radar. Cold fronts will often force whole flocks of migrating birds to drop to the ground and wait out the storm. These "fallouts" can force whole flocks to visit your yard until the storm passes. Migrating birds are grateful for high-energy food, so make sure to put out plenty of food and suet for them!

Heat

Birds generally have a higher tolerance for heat than they have for cold. Like us, during the heat of summer, they take afternoon siestas, reserving their activities for early and late in the day when it is cooler. Birds don't have sweat glands so they disperse heat by panting. On a hot day, you might also notice them raising their wings. They do this to circulate air around exposed skin. And, just like us, nothing tastes better for birds than fruit and berries and lots of fresh water on a hot day.

Cold

Birds have a couple of remarkable mechanisms for staying warm: First is the construction of their feathers. Birds are able to fluff up their feathers to create pockets of air, which add an extra layer of insulation around their bodies. Their soft, fluffy down layer is particularly useful in keeping them bundled up on a cold day. Secondly, birds have an amazing ability to slow down their metabolic rate and lower their body temperature. This helps them to conserve energy and stay warm in freezing conditions, but it uses a lot of their own energy, so it is important for them to replenish their energy stores after a cold snap.

Keeping a Calendar

If you want inspiration for keeping a calendar or a journal, you can look to no better source than Thomas Jefferson. Jefferson kept over sixty years' worth of daily notes. He kept records of the weather, when the first seeds

LEFT: A green heron lifts its feathers to cool itself off on a hot day.

LEFT: Instant down comforter? Mourning doves plump themselves up against the chill of a winter day.

were planted and harvested, and the arrival of each bird. Two hundred years later, his notes are still fascinating, particularly for those of us who live near his home in Charlottesville, Virginia.

To understand the pace and seasonal changes of your area, nothing compares to your own notes. I do not have the tenacity of Mr. Jefferson, and I realized long ago that I would not leave behind copious records of every change in the seasons. But it doesn't take many seasons of bird-watching to realize that you wish you could remember when the warblers came last spring, or when the nesting geese arrived at the lake, or when you saw the last hummer. So I keep a journal handy at the back door for jotting down the first sighting of a spring migrant, how many and what kinds of birds are at the feeders, and where they are nesting. Because my husband is a photographer, we have lots of photos, and often attach a feeder picture to a calendar page to capture the variety seen at the feeders that month.

I never fail to be enlightened and reeducated by looking back at my notes from previous years. Along with comments on joyful spring arrivals and the birth of fledglings, or the day that the cherry trees get picked clean every year, my journal also holds personal tragedies and failures: a dead sparrow in a nest box, a favorite wren's nest marauded by a snake, the lack of birdsong in an area where we took down diseased pine trees. These notes remind me of the interconnectedness of our lives with those who share the forest, and make me more aware of how my actions impact others.

> **"** If it looks like a duck, and quacks like a duck, we have at least to consider the possibility that we have a small aquatic bird of the family Anatidae on our hands. **"**
>
> **—DOUGLAS ADAMS, author of** *The Hitchhiker's Guide to the Galaxy*

②

Identifying the Bird

IT DOESN'T take a lot of bird-watching before you start asking yourself, *Hey, what's that one?* or, *Isn't that different from the other one?* Before you know it, you are scrambling for the field guide to name your find.

The first thing I do when I see a bird is note where it is. Is it in a marsh or in the forest? Is it high in the trees or sitting on a fence post? And what is it doing? Is it at the seed feeder or pecking on the ground? Is it cruising the skies or hopping around your shrubs? Noticing these things put you on the path toward identification. When you have basic information about bird behavior, you can use it to place a bird into its correct category. Before you know it, you can readily see that even though it is cute and tiny and brown, if you saw it hopping down the tree trunk, it's not a brown creeper. It's probably in the nuthatch family. That narrows things down considerably.

Taxonomy

I had the good fortune of being force-fed bird taxonomy in high school, so the basics were drilled into me long ago. Every bird has its order, family, and

LEFT: American goldfinches and one common redpoll.

RIGHT: If it looks like a duck . . . then you've made a start. Fortunately, this mallard is one of the easiest members of the duck family to identify.

genus, and I think I knew them all, at least at one time. The benefit of this knowledge, of course, is that you can recognize families of birds by reviewing a catalog of their physical characteristics, migration patterns, feeding behavior, and song patterns.

The first step to learning about a family is to study their silhouette. Read the description of the bird and then examine its general shape. Look closely at the body shape, type of bill, and tail length. Note anything distinctive about its posture. To commit it to memory, try drawing the silhouette yourself.

Then learn a few of the family's distinctive behaviors. Thrushes, for example, run or hop across the ground in a way that can be readily identified. Warblers look a lot like vireos, but vireos will perch for a short time, while warblers move incessantly. Knowing this can help you quickly reduce your choices from a field of candidates. If it looks kind of like a warbler, but is actually giving you the time to examine it, it probably isn't!

From the standpoint of the bird-watching hobbyist, the study of taxonomy can be focused on a handful of families. Most of the birds you are going to see in your backyard are Passeriformes, so this is a good place to start. These are the songbirds and perching birds, and there are over four hundred species in the United States alone. More than half of all bird species fall into this category. There are twenty-four different families in this group. Some contain birds that are readily identifiable. Nuthatches, cardinals, and chickadees fall into families that you may already know. Still others have limited representation in the United States. But other larger passerine families can be easy to confuse, and are worth taking the time to get to know on paper.

The following is a quick guide to some of these most common and sometimes confusing passerine families. These are generalities, of course. Like all families, variations inevitably occur!

Tyrant Flycatchers (Tyrannidae)

As the name implies, this family likes flying insects, so many of its characteristics are related to its ability to catch them. Its bill is pointed and has a small hook to help it snatch bugs from the air. Its demeanor is still and watchful—it has to be still in order to hone in on its next target.

- **Size:** Small to medium in size
- **Shape:** Large head with a high, rounded crown, strong substantial bill, short tail
- **Attitude:** Upright, sits very still, perhaps only moving head; holds tail straight down
- **Behavior:** Prefers insects; watch them dart out for flies and then return back to the same perch
- **Common family members:** Flycatchers, phoebes, eastern kingbird, and wood peewees

ABOVE: Flycatcher silhouette

LEFT: The eastern phoebe, a member of the flycatcher family, can be readily identified by his *fee-bee, fee-bee* call.

RIGHT: The regal golden-crowned kinglet is barely larger than a hummingbird.

ABOVE: Kinglet silhouette

Kinglets (Regulidae)

His name means "small king," and he indeed reigns in the forest as one of its smallest members. This family eats insects and insect eggs, so its needs to be agile and acrobatic to forage off leaves and small branches. Watch for it to feed this way, that way, even upside down.

- **Size:** Very small, the smallest of the passerines
- **Shape:** Short body with small, needle-like bill and an incised tail
- **Attitude:** Can be quickly identified by the constant flicking of its wings
- **Behavior:** A very nervous little bird, remarkably acrobatic and in constant motion
- **Common family members:** ruby-crowned and golden-crowned kinglets

ABOVE: One of the most identifiable members of the thrush family, the American robin has an air of confident authority.

Thrushes (Turdidae)

This rather large family is readily identified with its most well-known member, the American robin. Thrushes are generally ground foragers and have a pretty varied diet that includes earthworms and other insects, but they will also dine on fruit and berries. Take note that even though the bluebird is a member of this family, it has a slightly different posture and feeding style.

ABOVE: Thrush silhouette

- **Size:** Medium in size
- **Shape:** Round-bodied with a rounded head, strong chest, and a strong multipurpose bill
- **Attitude:** Confident and alert, with an upright posture
- **Behavior:** Runs or hops along the ground. If ground foraging is interrupted, it generally flushes to a tree branch where it remains quietly until it can go back to what it was doing.
- **Common family members:** American robins, bluebirds, and thrushes

Sparrows (Emberizidae)

Sparrows' plumage is generally muted in color. Shades of brown, gray, black, and rust dominate. They can easily be mistaken in the field for a finch or female birds of other species, so learn their shape and habits to help your identification efforts.

- **Size:** Small to medium
- **Shape:** Plump, round birds with short, rounded tails and small powerful beaks for crushing seeds
- **Attitude:** May cling to a plant stalk while it eats, even on the ground; sits upright in tree branches
- **Behavior:** Ground foragers, they can be seen scratching or hopping along the ground, looking for seeds or insects.
- **Common family members:** Sparrows, towhees, and juncos

ABOVE: Sparrow silhouette

Finches (Fringillidae)

Bright colors and brilliant songs—the finch family is definitely worth attracting to the backyard! Mostly vegetarian, they seek out the tiny thistle (nyjer) seed that other birds often ignore.

ABOVE: Finch silhouette

- **Size:** Small to medium-large
- **Shape:** Large round head, compact body, strong conical-shaped bill, and a notched tail
- **Attitude:** A generally gregarious family—look for them to feed in congenial flocks. Largely vegetarian, they seek out thistle seeds but will eat other seeds.
- **Behavior:** A bouncy flight pattern with a combination of flapping and gliding; they often call while in flight.
- **Common family members:** Finches, grosbeaks, redpolls, pine siskins

ABOVE: Finches thrive on a diet of tiny thistle seeds.

Wrens (Troglodytidae)

Seldom still, you might first notice a little wren because it seems to be everywhere at once, checking everything out. Very comfortable with suburban conditions, it will often make its nest in a planter near your house rather than out in the woods.

ABOVE: Wren silhouette

- **Size:** Small
- **Shape:** Compact bird with an erect tail that is easy to identify; a flat head shape; and a long, thin bill with a slight curve that lets it pick insects out of holes and crevices
- **Attitude:** Jaunty and upright, with that distinctive tail
- **Behavior:** An energetic demeanor that is more "busy" than "nervous"; often seen hopping quickly around low shrubs and thickets
- **Common family members:** Wrens

BELOW: The Carolina wren's loud *teakettle, teakettle* song is easy to recognize.

Wood Warblers (Parulidae)

Sometimes referred to as "the butterflies of the bird world," *both because of their tendency to flit nervously and because* *of their bright plumage. The biggest treat of springtime for* *those living in the East and Midwest, these small, colorful* *birds migrate en masse, sometimes arriving together by* *the hundreds.*

ABOVE: Warbler silhouette

- **Size:** Generally smaller than sparrows
- **Shape:** Petite and round, with a thin needle-pointed bill that can easily snatch insects. It's all about color with the warbler—most have distinctive markings, at least some in yellow.
- **Attitude:** Too busy to perch for long, can be found hanging upside down, checking out the underside of a leaf
- **Behavior:** Nervous energy—they tend to dart constantly from branch to branch
- **Common family members:** Warblers, chats, yellowthroats

ABOVE: A thrilling sight for fall birders is the migration of hordes of yellow-rumped warblers making their way to their winter homes.

Vireos (Vireonidae)

Easily mistaken for a warbler, these birds generally appear in gray, olive, and yellow, many with strong eyebrow lines and eye rings. The best way to distinguish them from the warbler is to watch their behavior. The vireo moves more slowly and patiently in search of its insects.

ABOVE: Vireo silhouette

- **Size:** Small
- **Shape:** Slightly larger than a warbler, with a heavier, hooked bill
- **Attitude:** Movements are deliberate—sits still and looks over a location before moving on
- **Behavior:** More solitary than the warbler; will travel in smaller mixed flocks
- **Common family members:** Vireos

Many other common families of birds may live near you. Woodpeckers, hummingbirds, owls, and hawks all have family traits that you may want to get to know better.

ABOVE: A cute little bird, but what kind? This Philadelphia vireo is very similar to both the warbling vireo and the Tennessee warbler.

Topography

Okay, so taxonomy may be useful in helping you narrow things down, but what we really want to know is, "What is *that* bird?" That's where topography comes in. Ornithologists have broken down every aspect of the bird's physical appearance into unique parts. This method allows you to understand how a bird's body is constructed, and helps you train your eye to go to the relevant markers that will help you identify it out in the field.

In the field, scan the bird from head to tail. Start with the big questions:

Size

How big is the bird? This can be a surprisingly difficult question to answer. We usually find ourselves comparing it to something we know really well. "It's bigger than a robin." "It's about the size of a titmouse." Sometimes its surroundings can help. We know the downy woodpecker from the hairy woodpecker because when it clings to our 5-inch suet feeder, we can tell exactly how long it is.

ABOVE: This downy woodpecker obliged our curiosity about its size by perching on a 5-inch suet cage.

Silhouette and Behavior

What is the shape of the bird? Does it flick its tail or is it still? Is it calmly perching or nervously fussing about? Is it on the ground or in a tree? Watch for attributes like the way a bird moves on the ground, or the shape and speed of its flight pattern. What season is it, and what's it do-

RIGHT: Despite seeing the distant flash of bright red, I wasn't sure I was seeing a cardinal. But when he got closer, it was obvious, even with his bald head.

ing? Birders call these traits a bird's GISS (pronounced *jizz*), shorthand for "general impression of shape and size," a quick holistic impression of the bird. As you get to know common birds in your area, you will get a feel for their *jizz*. With time and practice, you will find yourself putting them in the right category pretty quickly.

My sense of *jizz* triggered an alarm out in the field the other day. I was walking in our local nature preserve when my eye caught and held onto a flash of red. A cardinal! But no, something was wrong. The silhouette just wasn't right. Was it a scarlet tanager? I quickly thought of other red birds it could be, but nothing else made sense. I got out my binoculars and looked closer. Sure enough, there was his distinctive black mask. It *was* a cardinal. The problem was that he was bald. His crest was completely missing, and he had short stubble of black feathers on his crown. Poor guy; this happens sometimes during molting. But my brain is so accustomed to seeing that crest that my overall first impression of him had totally rejected the *idea* of a cardinal. Even with his brilliant color, that smooth crown made me want to seek another answer.

Flight

Identifying birds at a great distance isn't easy. I am not great with binoculars, so asking me to describe the details of a fast-moving bird is pretty futile. But even I can learn enough from a bird's flight pattern and wing shape to put them in the right family. There are some very basic wing shapes and flying styles that can help you quickly put a bird into the proper category.

Birds with **short, round-ed wings** are good at maneuvering through tight spaces

RIGHT: The distinctive wingspan of this hunting hawk is a common sight over open areas.

and operating in dense brush. Most of the songbirds are in this category, as are pheasant and partridge.

Flat, pointed wings don't allow for as fast a takeoff, but they are good for gliding and long-distance travel, so ducks and shorebirds tend to have wings like this.

Soaring birds, like hawks and other raptors, have **short, broad wings** that allow them to ride the air currents in search of food.

Large, arched wings create a long, slow flapping movement—hard to miss in the egret or heron families.

Lastly, the amazing **hummingbird's wings** rotate at the shoulder, allowing it to execute a figure-eight movement for nearly motionless hovering, to sip nectar.

Once you can put the bird into the right family based on its wing shape and flight style, it is time to study it closer so you can take the next step of identifying its species.

The Bird: An Anatomy Lesson

Now that you have formed a general impression of the size and shape of your bird, it's time to get down to the details.

The Bill

The bird's bill says a lot about it. The shape will give you a good idea about what it eats. **Short, triangular bills** are good for crushing seeds. **Curved or pointed bills** are good for snatching insects. If you see a **long bill,** that

LEFT: The wren's long, curved beak allows it to reach into small places to snatch an insect.

RIGHT: The northern cardinal's short, thick bill is ideal for crushing seeds.

RIGHT: The great egret's long, straight bill makes him well suited for fishing.

BELOW: The eastern bluebird's short, straight bill is equally suited for snatching insects and picking fruit. No sunflower seeds for him!

bird may catch fish for its meals. Because some birds are generalists, their bills may be a combination of traits and ideal for eating a wide variety of foods. Take note of bill shape and length. Color varies also, and can be a quick identifying marker. The cardinal has a red-orange bill. The dark-eyed junco's bill is pale and pink, one quick way to realize you are not looking at a phoebe.

Color and Field Markings

Quick! What color was that bird? Try to capture your overall impression. Was it bluish-gray, fading to white? Was it brown—and if so, what kind of brown? If you saw a splash of yellow or blue or red, where was it? Did it have wing bars, an eye ring, or a stripe? What color was its belly? Take note of everything you see, but don't be too quick to identify your bird yet, unless its color is very obvious. Seasonal variations, gender differences, and juvenile markings can all quickly confuse your ID. So keep looking!

The Head

That little head offers a lot to look at and may hold the key to your bird's identity. First, look at the **crown.** The crown is the top of the head; look for patterns of striping there. **Median stripes** go down the center of the head. **Lateral stripes** follow along the sides of the crown. Some birds have a **crown**

ABOVE: Just a little brown bird? This white-throated sparrow's multiple field markings can tell you a lot about him.

patch. This is a contrasting color at the top of the head. Although distinctive, it can be hard to see. The ruby-crowned kinglet is a good example. Its splash of color is hidden by its other crown feathers. The most obvious crown feature is the **crest,** a peak of longer feathers which can be rigid and upright or softer and more relaxed.

The rest of the head provides several different markings. **Eyebrow stripes** are over the eye, while **eye lines** go through the eye. **Whisker marks,** also known as the mustache or malar stripe, is just under the eye. The **throat patch** is right under the bill. The presence of an **eye ring** can also help you get a positive identification. Check out the eye color, too, if you have a good-enough view.

Underside

When we talk about a bird's underside, we generally are referring to everything from its chin and throat to its breast and belly. Note the color of each and whether there are any distinguishing

RIGHT: The eastern towhee's white belly quickly distinguishes it from a robin or a redstart.

ABOVE: A closer look at the white-throated sparrow shows an obvious eye stripe, crown color, and throat patch, as well as the yellowmarking on the side of his head.

marks. In some species, this may be a key element. Both sexes of a species may share a similar trait, like a white belly, in spite of other differences.

Upper Side

The back of the head, the collar, the mantle, and the rump make up a bird's upper side. Note the color and patterning of each. This is the part of the bird that generally provides the initial sense of his overall color.

Wings

Wing color and pattern are distinctive in many species. Some birds, like warblers, have strong **wing-bar patterns.** These are stripes across the folded wing. Still other birds, like the red-winged blackbird, have a strong identifying block of color called the **shoulder patch.** Other times, the underside of the wing

BELOW: Note the wing shape and long primary feathers of this vulture, as well as the secondary-feather field markings.

may tell the tale. The wing lining, which are the feathers covering the underside of the wing, may be a different color or pattern. There are long **primary feathers** on the outer half of the wing and shorter **secondary feathers** on the inner half. The **speculum** is a patch of colored secondary feathers that help to identify many ducks. Look for the iridescent green speculum on a green-winged teal, or the iridescent purple-blue with white edges of the mallard. **Wing tips** may show some white.

Tail

Notched, pointed, square, rounded, or forked? Long or short? Up or down?

BELOW: A common European garden bird, the blue tit has a blue nape, wings, and tail, with a yellowish-green back.

Flicking or still? Note the tail's markings, and whether there is a white tip or any white patches above or below. Also note what it is doing. Some birds spread their tails out as they settle on a branch. Still others flick their tails up and down. Some woodpeckers use their tails as a brace for climbing.

Now that you have made a mental note of all your bird's traits, jot down

your impressions, or do a quick sketch. If you think you have placed it in a certain family, make a note of that too. Because my husband and I often have a camera along, we will try to snap a quick photo, even if we know it won't be great quality. After we are home and have had time to reflect on what we saw, we get out our field guide and compare notes. Having the time to really look and study is a great way for us to examine the possibilities and make inferences.

LEFT: This bird's forked tail immediately places it in the swallow family.

Variations

Sometimes, even with good notes, a bird's identification may be elusive. If birds were completely predictable, what fun would that be? While it may seem like enough work for the novice bird-watcher to figure out the differences between all the birds out there, the addition of age, gender, seasons, and other factors complicate things further. Some birds, like the blue jay, may never fail an ID test, but many others may show differences in color and markings that make identification difficult. Keep in mind that generalities are dangerous when making identifications—nearly every rule seems to have an exception!

Gender

Some bird species are monomorphic, meaning the two genders have the same general size, shape, and plumage. But more common, especially among songbirds, are dimorphic differences—males and females of the species having different appearances. These differences can be subtle or significant.

Monomorphic Birds

Black-capped chickadees are virtually indistinguishable from one another, as are blue jays and cedar waxwings. Owls, crows, and titmice are among others that are nearly impossible to tell apart, although size may be an indicator. Most males of the species are larger, although that is not always the case. Female owls are somewhat bigger, as are eagles and some shorebirds.

Close observation of a pair together may help you to capture subtle differences, but looking at their behavior may be the best way to make an educated guess. Males usually migrate before females, so if a new species appears at the feeder, it may very well be a male. Males are the real singers of the family, too, so generally it is safe to say that a singing bird is the male of the pair.

During mating and nesting season, differences may be obvious. During mating, males will be much showier, and you may see courtship rituals such as flight and display patterns, singing displays, and even courtship feeding of the female. During actual mating, the male is always on top. Nesting habits may also be a tip-off. Although different species have different habits, you may be able to spot the female because she is doing all the nest building or tending.

My Carolina wrens are relatively easy to tell apart in the spring—at least, some of the time. He is the singer in the family, so the distinctive *teakettle, teakettle* will be his. She tends the eggs alone, though, so before the eggs hatch, she is the only one there.

Dimorphic Birds

This category includes subtle variations, like the muted color difference of the American robin female, to the riot of variations in the wood warbler family.

Most gender variations are all about plumage color. A northern cardinal female, for example, displays the same black face, orange-red bill, and sharp crest as the male, so she is generally easy to spot. She will be brown with reddish highlights. (Okay, I'll admit it: I think she is actually prettier than the male!) Likewise, the eastern bluebird female has just tinges of her partner's blue tones and a softer, rusty breast. Still, she is pretty easy to spot.

Have you ever looked at a human couple and asked yourself, "What are those two doing together?" It's sometimes like that in the bird world, too. The male and female in species like the red-winged blackbird are completely different from one another. He is black and flashy with red shoulder epau-

BELOW: This blue jay carries identical markings to his mate's.

lets. She is a muted brown bird with a streaked belly and a much smaller size. Seen alone, you would probably think she was in the sparrow family. Likewise, the rose-breasted grosbeak; his crisp black, white, and red markings are impossible to miss, while she is the picture of understated elegance—soft and brown, with subtle markings and streaks that also make you think of the sparrow family. The best way to identify these birds is during the mating season. If you see a male and then notice what appears to be its partner, it will give you the opportunity to examine and get to know her on her own.

Age

Look—a singing female! No; unless it's a cardinal, you have probably just seen an immature male. In the bird world, a lot of babies start out life looking like Mom. Juvenile songbirds wear their "subadult" plumage for as long as a year. Some larger bird families take even longer—three to five years for an adult bald eagle. Until then, it might be easy to mistake an immature bald eagle for a golden eagle.

Molting

And then there's molting! While birders out in the field may obsess over the intricacies of the complex molting process and the changes in the appearance of common species, for the purposes of the bird hobbyist, it is simply important to know that it does happen. Molting is the process of shedding and replacing feathers. Given the wear and tear on feathers from weather, flight, nest building, and defending territories, it's no wonder that birds need a regular spruce-up. For most birds, this process happens twice a year.

After breeding season, a complete molt generally occurs. While this means the replacement of all the feathers, the process actually occurs in an orderly and subtle manner. If that were not the case, you would see a lot of naked birds around! Instead, what you may see is a bird that literally looks like it's trying to grow out a bad haircut. Old feathers are replaced with small pin feathers, which need time to grow. Molting birds look unkempt and uneven, with color variations that may give you pause. Many flashy males will have a softer, more-subtle look after molting. In many species, they spend the fall with an appearance that is more like the female.

For some birds, like monomorphic chickadees, owls, and others, one complete molt per year seems to be all that is necessary. Still other birds that live in harsher weather and environmental conditions may have two complete molts every year.

In some species, another molt occurs before mating season. This is only a partial molt, and is often referred to as the prenuptial molt—a fitting name for putting on your fancy duds in preparation for going courting!

ABOVE & RIGHT: The flashy red male northern cardinal (above) gets most of the attention, but I have always been partial to the understated sophistication of the female's muted colors (right).

This is when many songbirds put on their brightest colors. The American goldfinch is a good example. Yellowish-green and subdued like his pals all winter long, his prenuptial molt reveals a glorious, flashy little canary-like bird. How can any female resist?

Replacing all one's feathers takes a lot of energy. Your backyard birds will do this right after breeding season, when food is still abundant and the stresses of rearing their young are behind them. They will usually complete their partial molt before leaving their winter grounds so that their feathers are in top condition for the long trip.

Seasonal Changes

The molt that happened in midsummer has produced a change in many of your favorite visitors. Spring brings a riot of color—bright yellow warblers, hot, spicy orioles and tanagers, and brilliant bluebirds. In the fall, though,

BELOW: About to molt? This little kinglet's feathers are all shook up!

these bright and flashy males are in their "off-season" garb, often resembling a softer version of their former selves. It is often harder to distinguish males from females in the fall. Scarlet tanagers are a good example. Unmistakable in the spring, in the fall the male softens to the olive-green color of the female, with only hints of his brilliant red past. Youngsters are growing to adult size, but may be somewhere in between their juvenile and adult plumage.

Warblers are considered notoriously difficult to distinguish during the fall, but this may be a slight overstatement. While all fall warblers will appear in fresh molt and may be a little more subtle than they were in the spring, their markings are basically the same. If you see what you think are a lot of females together, or even a lot of males in dull attire, what you may be seeing is a lot of immature birds. If you are really interested in ID'ing your warblers, concentrate on shape and pattern. Wing bars and face patterns should still be identifiable, and individual calls can also help.

Unusual Plumage

There are a few conditions that cause variation in the normal appearance of birds. One is albinism—a complete lack of pigment that causes a white appearance and pink eyes and extremities. A variation of this condition, called leucism, may produce only a partial loss of pigment, usually with normal eyes and extremities. Although these are rare conditions, you may catch sight of a robin or other common bird with this unusual plumage. There are also conditions that produce increased dark color, increased yellow, and even a confusing mix of male and female plumage—all uncommon, of course, but interesting to know.

Rare Bird Sightings

If you are having trouble identifying a particular bird, there is a slim chance that it is in fact an unusual find, and not just some variation of the local population. Experienced birders sometimes identify species that may be common in other areas, but are way outside of their normal range. Likewise, some birds are pretty rare, and the thrill of spotting one may be lost in your anxiety to make sure it is what you think it is. In cases like this, try to get a photo or a second pair of eyes to help you confirm that you are seeing what you think you are.

ABOVE: This American goldfinch is in his "off-season" garb. In spring, he will be a dazzling yellow.

Field Guides

The classic guide for bird identification is *Peterson's Field Guide to Birds of North America*, which is organized by bird orders and families. It's the one I grew up with, so I am fairly loyal to it.

If you buy a regional guide, choose one for your entire region rather than just for your state. Make sure range maps are included, as these will help you to determine whether a bird is a seasonal resident or just passing through. Regional guides are handy because they are not as bulky as the more-complete guides, but don't assume that they are all as complete as they should be. A friend of mine once photographed a golden-crowned kinglet and called me excitedly about her rare find. Because she had not seen it in her regional guide, she assumed it was unusual. In fact, golden-crowned kinglets are fairly common around here, which she realized as soon as she consulted a more-comprehensive guide.

Some guides use paintings, while others use photographs. The rationale for paintings is that they can capture the bird in the perfect posture to highlight key field markings. Photographs have to contend with lighting issues and variations, as well as the challenge of capturing an angle that will show off the bird's key features. I appreciate both styles, and sometimes switch back and forth when learning about a bird. Photographs give me hints about seasonal and gender variations that the paintings sometimes do not reveal.

Choose your guide based on the type of organization you are most comfortable with. Some books are organized strictly by bird families, while others are organized by color, or have helpful tabbed color references.

I have "bookshelf" guides—references that I go to at home for a complete description and comparison of similar birds. Then I have "field" guides—smaller, pocket-sized guides that are durable enough to withstand being shoved into pockets and backpacks and taken out during bad conditions. If you are serious about birding, before you know it, you will own more than one!

See the appendix at the back of this book for a list of the most-popular field guides.

An eastern bluebird pair

" In order to see birds it is necessary to become part of the silence. **"**
—ROBERT LYND

③

25 Backyard Birds You Should Know

NOW THAT you know how birds live and the basic elements of identification, it is time to get to know your own neighbors more intimately. You may have seen many of these birds, and even if you are not an experienced birder, you could easily read through this list and say "Been there, done that." Everyone knows a robin, after all, and who can't say they have seen a blue jay? The question is, do you know anything about them?

Take the time to read about the lifestyles of some of your backyard neighbors, and learn to identify the common ones whose names you don't know. Learning the intimate details of a robin's life will help you identify any member of the thrush family more readily. And understanding the trials and tribulations of the chickadee will make you really appreciate the tenacity of this common little bird. Knowing about your most common visitors lays the groundwork for all of your birding skills.

Red-Winged Blackbird

Anyone who pays attention to birds can't fail to notice the red-winged black-bird. A stocky little bird, it's a bit smaller than a robin, with a slender bill and a medium-length tail. As is the case with a number of bird species, it's the male that gets all the attention. Male red-winged blackbirds are glossy black with patches of red and yellow on their shoulders. He is hard to miss, because he tends to be an incredible show-off, particularly during mating season. A male red-winged blackbird has a cheerful song that he readily shares. He sits high on perches and performs his loud and distinctive *conk-a-ree* all day long. There is a reason for this exceptionally showy behavior. The red-winged blackbird is a polygamous species, which means the male needs to attract a lot of females; he may have as many as ten mates in his territory. Mating and breeding season is a very busy time for him. He is an aggressive defender of his turf, and it's not unusual to see him go after much larger birds in his quest to keep his dominion safe.

Females are much more subdued. They are a dark streaky brown and can be mistaken for a large sparrow. The female tends to stick with more-domestic habits, staying lower to the ground, foraging for food, and working on her nest. Her typical sound is *chit-chit-chit-cheer-cheer-teer*. She generally breeds in marshes, although it's not unusual for her to choose to nest in a pasture or a wooded area near a river. For safety's sake, she prefers to join a small colony of other females during breeding season. Although black-birds raise two or three broods per season, they are meticulous parents, building a new nest for each brood. Babies tend to look like Mom at first, with the males' colors blossoming out as they mature.

Red-winged blackbirds prefer wet areas, marshes, and small watering holes much of the year, although you can also see them together in fields and pastures. They are a social

RIGHT: Red-winged blackbird

bunch, often congregating in flocks, called "clouds," during the winter months, frequently with other birds such as grackles and starlings.

Red-winged blackbirds are omnivorous. Their diet consists mostly of a variety of seeds, as well as corn, wheat, and other grains. About 25 percent of their diet consists of protein: They eat insects in the summer by picking them from plants or catching them in flight. Of course, they also enjoy summer berries.

Backyard feeders can offer sunflowers and other seed mixtures, including suet, to attract red-winged blackbirds. But plan for a large party! Autumn and winter visitors travel in large flocks, often numbering into the hundreds.

Eastern Bluebird

This small thrush is a darling little bird with a big, rounded head, large eyes, and a short, straight bill. He has a plump body with a distinctive white

ABOVE: Eastern bluebird

belly and a perpetually alert posture on his short legs. It is no wonder his likeness is a favorite feature in Disney movies!

The male eastern bluebird has a rich blue back, with a rusty red throat and breast. He has a short *chir-wi* warble, as well as a song that sounds like *chitti wew we we do.* He makes a great show of being a homeowner, attracting females to his chosen nesting location by carrying nesting materials in and out of the hole and posturing at the entrance by fluttering his wings. Once he has selected a space, he is fierce in its defense, attacking any birds he considers a threat, whether they are cavity-dwellers or not. That's pretty much his entire contribution to the brooding process, however. Once he attracts a mate, he leaves the rest of the nest building to her. He tends to be a one-woman bird; when a pairing has been established, a bluebird couple will often stay together for several seasons.

Female eastern bluebirds have similar markings to the male, but their colors are more subdued. The female's back is an elegant gray with tinges of blue on wings and tail. Although her breast is a softer orange-brown than the male's, her white belly is just as easy to spot. After the male has chosen the nesting site (usually a natural cavity in a dead tree, or a hole left behind by a woodpecker), the female weaves together her nest using grasses and pine needles. She may have more than one brood each year. An early brood will leave their parents during the summer; however, if she has a later brood, they may choose to winter with their parents. The female will use the same nest for both of her broods.

Eastern bluebirds are fun to watch for, because you can often spot them perching alertly on wires or fence posts in open spaces. They like open country with sparse ground cover for hunting, so human environments such as parks, backyards, and golf courses are attractive to them.

Eastern bluebirds have keen eyesight—a trait they need for catching sight of their insect prey and dropping to the ground to grab them. Most of their diet (approximately two-thirds) is insect-based. They eat grasshoppers, crickets, and beetles, but will also indulge in earthworms, spiders, and snails. Fruits and seeds from shrubs like dogwood and hawthorn are an important part of their fall and winter diet.

Eastern bluebirds don't frequent conventional bird feeders, although they are happy to oblige if you choose to offer live mealworms. They are also attracted to birdbaths. The best way to bring them to your yard is by offering nest boxes. Because competition for tree cavities is fierce among species, a nest box can be a welcome option.

Northern Cardinal

We have several cardinal pairs in our yard in Virginia, and we never fail to thrill at the sight of them. They are the quintessential songbird—a showy bird with a long tail, a short, thick bill, and a prominent crest that is easy to spot.

ABOVE: Northern cardinal

The male northern cardinal is, of course, the star of this show. His brilliant red color is a captivating sight, even from a distance. At close range, he is even more handsome, with a dapper black mask around his face. He tends to seek high branches to sing from. His song is a loud whistling *cheer-cheer-cheer-what-what-what*, but you might also be able to quickly identify his common warning call, which is a sharp *chip*. Male cardinals are obsessed with defending their territory during breeding season, and stories abound of cardinals fighting their own reflections in car mirrors or windows. Pairs usually forage together during the breeding season, but you will often see them in larger groups later in the year and into the winter. Pairs sometimes stay together for the next season.

The female is a warm brown with a pretty reddish tint, and because she carries the same sharp crest and black face as the male, she is easy to identify. The female northern cardinal is unusual among songbirds in that she actually sings, even while sitting on the nest. She and her mate take an egalitarian approach to the selection of their homesite, often visiting several sites together before making their final choice. The northern cardinal's nest is an elegant thing. After selecting the crook of a tree, preferably behind a screen of dense foliage, the female uses small twigs twisted into a round cup shape, and then lines it with layers of softer materials. As beautiful a feat of engineering as it is, she generally only uses it once.

Northern cardinals are common in backyards, parks, and along the fringes of the forest. They like dense tangles of shrubs and trees, so landscaping shrubbery is often attractive to them. They do not migrate, so they are especially fun to catch sight of during the winter. It is no wonder that holiday greeting cards often feature a brilliant red cardinal on a snowy background!

The diet of northern cardinals consists primarily of seeds and fruit. Dogwood, wild grape, buckwheat, mulberries, and blackberries are all natural attractors. They drink maple sap from holes made by other birds. Insects such as beetles, crickets, and grasshoppers are an important part of their spring diet, as they feed their young meals almost entirely made up of protein.

Cardinals are easy to attract to the backyard feeder. While they eat many kinds of birdseed, standard black-oil sunflower seeds are their seed of choice. They are a joyful addition to the winter feeder scene, and often happily forage with other species.

Black-Capped Chickadee

If we were conducting a popularity contest of the bird world, chances are the hands-down winner would be the black-capped chickadee. Who can resist

that adorable round body, the alert curiosity, and the apparently perpetually cheerful nature?

The chickadee is easy to identify by a large head on top of a round little
body. He has a long, narrow tail and short beak, which quickly differenti-
ates him from the nuthatch that beginning birders occasionally confuse
him with. The black-capped chickadee wears a dapper black cap and bib
with white cheeks. His back, wings, and tail are gray, and he has a whitish
belly with buff-colored sides. Males and females have nearly identical
appearances.

The male black-capped chickadee is a congenial mate; he helps the
female excavate the nesting area where she will build their nest. He also
takes charge of most of the feeding duties until the female is done brooding.
Black-capped chickadees mate for life. The pair generally has one brood per
season.

Chickadees are not an especially wary bird, often associating with
woodpeckers, nuthatches, and warblers, and perfectly willing to investigate friendly humans. That being said, they do adhere to a fairly strict

ABOVE: Black-capped chickadee

social hierarchy. Offspring do not belong to the parents' flock, and visiting birds have a lower rank in the group. Black-capped chickadees rarely perch for long, but prefer to sweep in, grab a seed, and settle elsewhere to eat it. They are the hoarders of the bird world, hiding seeds in different spots for later use in the winter. And they have an extraordinary ability to remember where they put the seeds!

Both the male and the female are remarkable vocalists. Their calls are complex communications between individuals and flocks. They are known by the musical song for which they were named, the *chick-a-dee-dee-dee*, but you will also hear their common call of *bee-bay-bee-bay*. Many bird species that associate with chickadees will actually respond to a chickadee alarm call.

Black-capped chickadees are year-round residents of deciduous forests, although they may find it convenient to move from the forest to your backyard during the winter months. In the coldest weather, chickadees can reduce their body temperatures to conserve energy. Chickadees maintain a largely protein-based diet during most of the year, eating insects and spiders that they locate underneath leaves and bark. They feed their babies a high-protein diet, as well. In the winter, they add seeds and berries to their menu.

Black-capped chickadees are a joy at the feeder for several reasons. First, their natural affability and curiosity about humans often allow you to feed them right from your hand. They are easy to feed, and are often the first ones to find a new feeder. They like sunflower seeds, peanuts, suet, and peanut butter. Because of their sociability, they often bring along interesting visitors to your feeder that might not have stopped in on their own.

Blue Jay

Noisy, aggressive, and outspoken, it's no wonder that a flock of blue jays is often known as a "scold of jays." They are an intelligent and curious bird that will often watch humans fill a feeder, plant the garden, or walk away from the picnic table, and then sweep down when the coast is clear for an easy meal. They have a bad reputation within the bird community, as well, for behavior such as raiding other birds' nests and chasing other birds from the feeder.

This very common, large songbird has a prominent crest and blue-and-white plumage with a black collar. His crown of head feathers indicate his

ABOVE: Blue jay

mood: It may be flat to his head when he is relaxed, or raised and bristling if he is afraid or excited. The blue jay makes a variety of sounds, from his *jayer-jayer* call to his squeaky "rusty pump" sound. He is an excellent mimic, often crying like a red-shouldered hawk to scare away other birds.

Blue jays build their nests of roots, twigs, grass, bark, and paper. Nests are generally placed fairly high in the crooks or thick branches of either deciduous or pine trees, but they are not all that picky, and will make the best of any situation, sometimes simply taking over a handy robin's nest. Males share all the work of nesting, from helping to gather materials and assemble the nest to feeding the brooding female and fledglings. The sexes share a similar appearance.

Although fledgling blue jays are a soft gray color, they exhibit the distinctive wing pattern and head markings from an early age. They also exhibit the family's strongly adventurous spirit, occasionally wandering away from the nest a little earlier than they should. It is not uncommon to come upon one of these stray youngsters squawking his heart out. If he can be placed back in or near the nest, the parents will usually accept him back. The family generally stays together through the early fall. Blue jays are partially migratory. Some migrate out of wintry locations, while others choose to winter over.

Blue jays eat just about anything. They have a special fondness for acorns, but are known to eat seeds, grains, berries, and fruit, as well as insects and other small invertebrates. They are not above the occasional scrap of human food, either, and are happy to steal bread or meat from an open garbage can or picnic table. They can carry away quite a lot of food, storing some in a pouch in their throats while holding more in their mouths and bill. At the feeder, blue jays will come for peanuts, sunflower seeds, corn, and suet.

American Robin

The American robin is probably the most recognizable of America's backyard birds, with a range that covers all of North America. Even non-bird-watchers take note of this fellow's arrival in the spring, signaling the end of a long winter. In reality, in all but the northern states, American robins do not migrate at all—it is just that typical suburban lawns don't hold that much appeal outside of earthworm season!

The American robin is a fairly large, heavy-breasted bird with grayish-brown feathers and a brick-red breast. He has a strong, straight flight pattern that is low to the ground, and a habit of flicking his fairly long tail in a downward motion several times before settling on his perch. These traits make him easy to notice from a distance.

The classic "early bird," the American robin has the distinction of being early in a number of ways. An industrious worker, it is out and about by dawn, often breaking the morning silence with its lyrical song before other birds are up. It hunts early, too, hopping across lawns, looking for earthworms and other small insects, head tilted as though listening for its prey. The robin is also the earliest to lay its eggs, often establishing its nest as soon as it arrives in its summer range.

The male sings and spreads his tail during courtship, giving chase to any female that looks his way. He is a protective suitor. Once he has chosen

ABOVE: American robin

his female, he stays pretty close to ensure that another male doesn't horn in. There is some evidence that the male provides food to the female even during courtship—a way to keep her nearby and help her gain weight in preparation for the strenuous nesting season. Although it is sometimes thought that robins mate for life, this is not the case. American robins mate for an entire season, and though they may meet up and join again the following season, it is not a sure thing.

The female has the same markings as the male, but her colors are more muted. She will have as many as three broods in a season. She builds her nest of mud, twigs, grass, and feathers. In summer, females sleep on their nests. Males and young adults gather together on roosts. Both the male and the female feed their brood.

American robins prefer open woodlands, so parks, backyards, and suburban areas are preferred during spring and summer. Depending on where you live, they may winter over, but you won't see as much of them because they spend more time in large groups, roosting in trees, rather than foraging on your lawn. An area of woods with fruit trees and berry-producing shrubs makes an ideal winter home.

Robins are well-known consumers of earthworms, but they actually eat different types of food depending on the time of day. Earthworms tend to be their breakfast food, while they seek out fruits and berries later in the day. They feed their brood a combination of plant and animal food.

Robins are not particularly interested in feeders, although they might appreciate a snack of apples or bread on a cold or snowy day. They are ground foragers, so find a safe spot where they can get to your offerings. If you want to help them out during a late-season blizzard, they will also come to eat mealworms.

Ruby-Throated Hummingbird

I often hear a ruby-throated hummingbird before I see it. Beating its wings over seventy times per second, the *whir-whir* sound made by its acrobatic movements makes me quick to turn my attention to the nectar feeder.

A tiny jewel of a bird with an iridescent brilliant green back and crown and a grayish belly, it should come as no surprise that a group of humming-birds is often called a "shimmer." A full-grown bird is about three inches long and weighs less than two-tenths of an ounce. The distinctive ruby throat appears only on the male. Females and juveniles have a green back and grayish belly, sometimes with a black eye mask and a black tail with white spots. They are amazing aerial acrobats, with the ability to hover motionless in midair, and move up, down, sideways, and backwards with ease.

A ruby-throated hummingbird's heart beats 225 times a minute when the bird is at rest, and rises to more than 1,200 times per minute when it is in flight. Its wings beat about 70 times per second in direct flight and over 200 times per second while diving. It has a small mouse-like twitter that is generally only used for alarm calls.

Males have dramatic courtship displays, diving in dramatic U-shaped sweeps. The male stays around only for courtship and mating, leaving the female to tend the nest alone. Males aggressively defend the flowers and feeders in their territory, especially in late summer and early fall, as they prepare to begin their long nonstop trek south toward the Gulf of Mexico. The hunt for food is a constant one, and good feeding locations can draw major battles between birds.

The female builds a nest of delicate intricacy, using spider silk, thistle-down, and moss. Nests are placed on the slim branches of trees, usually 10 to 40 feet off the ground. She feeds her brood regurgitated insects.

ABOVE: Ruby-throated hummingbird

Ruby-throated hummingbirds prefer deciduous forests, old fields, and orchards, and are at home in suburban backyards. They are a delight to watch. Look for them in gardens, especially around red and orange flowers. You might also catch them working the yard, snatching small mosquitoes and gnats out of the sky. Males arrive first in the spring. Both males and females are on their way to their winter homes by late summer or early fall.

Ruby-throated hummingbirds feed on the nectar of red and orange flowers that have relatively high sugar content. Tubular flowers are favorites, allowing the hummers to use their long, tubelike tongues to reach the nectar. Hummingbirds hunt small insects, too, catching them in midair.

At the feeder, hummingbirds are easy to please, asking only for a drink of sugar water from a specially designed nectar feeder. A four-to-one ratio of purified water to sugar is all that is needed. Red dye is not necessary, and may actually be harmful. Depending on your area, you may have hummers into the early fall. Here in Virginia, we keep our nectar feeders out until early October. Another way to please hummers is to plant their favorite flowers. Trumpet creeper, cardinal flower, honeysuckle, bee balm, and red morning glory are all great hummingbird attractors.

Baltimore Oriole

Fresh oranges and grape jelly! Compared to other songbirds, the dining habits of this fellow may seem heavy on the sugar. While the Baltimore oriole's diet consists largely of insects and caterpillars, it is true that he seems to have a weakness for sweets.

The male Baltimore oriole is hard to miss. He is a little smaller than a robin, and has a brilliant orange underbody, shoulder, and rump. The rest of his body is black, with a single white wing bar.

ABOVE: Baltimore oriole

The male arrives in the spring a few days ahead of the female to claim its territory. He has a fluid, fluty song which is somewhat similar to a robin's. The Baltimore oriole will sing from dawn to dusk while searching for a mate, chattering and preening for her. Pairs mate for a whole season, and both males and females work to feed their brood.

The female Baltimore oriole has a yellowish-brown upper body with a dull orange breast and belly. Juvenile orioles also have this appearance, often into the second year. The female does all the nest building, crafting a tightly woven pendant nest made up of animal hair and fine grasses and fibers which she hangs on the end of a branch.

Baltimore orioles prefer open woodlands and light forests, making parks and suburban areas attractive locations for them. They tend to frequent the same area year after year, so even though attracting them to your yard may test your patience—all that wasted jelly!—in the end, you may be rewarded with yearly visitors.

Because Baltimore orioles eat a diet that consists primarily of insects and caterpillars, they can be a great addition to the backyard environment, especially if tent caterpillars decide to set up shop in your trees. Be careful not to use pesticides to control these pests, as they could sicken the birds that are trying to help control the population. Baltimore orioles also like fruits and nectar. The ones that visit us are devoted to our cherry trees—or should I say, *their* cherry trees! While they can be seen sipping from a humming-bird feeder, there are also specialized oriole feeders on the market that hold nectar, halved oranges, mealworms, or grape jelly.

Northern Mockingbird

Perched high on a tree, a fence, or a telephone wire, the subtle colors of the gray northern mockingbird belie his true nature. A talented and showy

ABOVE: Northern mockingbird

vocalist, he sings from dawn to dusk, and beyond. He can often even be heard at night. He has a large repertoire of songs, as well as the ability to mimic other birds. While the male is the primary singer, the female also sings.

A larger songbird, the northern mockingbird is approximately ten to eleven inches in length. He is longer and more slender than a robin, with a long tail that he may flick from side to side. He has short, broad wings with two white wing bars. A large white wing patch is visible on mockingbirds in flight. His tail is black with white outer tail feathers. Females are similar in appearance.

Northern mockingbirds have big personalities and are fiercely territorial, making them pretty entertaining to watch. They are happy to pick a fight with anyone who comes along intending to infringe upon their space, including themselves if they happen to catch a glimpse in a window or reflective surface. You might notice a mockingbird picking a fight with a dog or cat, and it may appear to defend a backyard feeder even if it doesn't feed there. They can bark like dogs, whistle like humans, and I am sure I have one that can imitate a ringing phone!

Male mockingbirds do a courtship "loop flight," which is a quick flight up, then a loop right back onto their perch. They combine this with constant singing until they attract a female—often into the night. When the nesting site is chosen, usually in a low shrub or tree, the male builds the twig base which the female will then finish and line. He helps to raise the young and may be the primary caregiver to the fledglings while she moves to a second nest to lay another clutch.

Because the northern mockingbird prefers open ground with shrubbery and thickets, it is well suited to parks, farmland, and backyards. Ground feeders like the robin, northern mockingbirds can be seen running or hopping on lawns, looking for earthworms and other delicacies. They enjoy fruit and berries, too, so even though they may not come to your feeder, they can be encouraged to visit your fruit and berry bushes. In most areas of the country, they are year-round residents. At the feeder, they may come to investigate suet or fresh fruit or raisins.

American Goldfinch

Sometimes referred to as "America's canary," this bright yellow little bird is a year-round visitor in most of the country. The male wears bright colors

ABOVE: American goldfinch

for spring mating season, dulling to a muted tone like the female in the off-season. Although his yellow color may immediately make you think "warbler," take a close look at its bill. Unlike the warblers and vireos that have insect-catching bills, the goldfinch is a strict vegetarian, and has a short conical bill which it uses for eating seeds.

Because its natural diet consists of weed and thistle seeds that mature later in the summer, American goldfinches tend to nest a little later than many other songbirds, raising their young in June or July.

Males perform elaborate singing and flight displays during courtship. Mated pairs choose their nest site together. The female builds the nest in a shrub using a weaving technique that produces a nest so tight that it actually holds water. While the female incubates the nest by herself, the male may help by bringing nesting materials or feeding her on the nest. Couples only raise one nest per season. American goldfinches prefer gardens, open country, and pastureland, where seeds are abundant, and they seem quite at home in proximity to human settlements.

The American goldfinch is easy to spot at the feeder, where you will see sociable flocks of goldfinches dining companionably with other finches, common redpolls, and pine siskins. Weed seed is their food of choice, so put out specially designed tube or sock feeders for holding nyjer seed.

Downy Woodpecker

The downy woodpecker is one of North America's most common woodpeckers, although the seeming abundance may be partly due to birders often confusing it with its look-alike cousin, the hairy woodpecker. The downy lives year-round across most of the country, a common sight at feeders, where it comes to visit the suet cage and to sample some seeds. It happily shares space with nuthatches, titmice, and chickadees.

A distinctive "little soldier" posture marks the identity of the downy woodpecker. An upright little creature, this bird has strong black-and-white markings and a distinctive red patch on the back of its head. At 6 to 7 inches long, it's a full 2 inches shorter than the hairy. If you were to compare the two side by side, you would also note that the downy's bill is somewhat smaller for its overall size. In the field, note its call. The downy woodpecker says *pick*, while the hairy has a sharper, harsher *peek* call. Likewise, their whinny-like trills are slightly different. Females have the same markings, but lack the red patch.

LEFT: Downy woodpecker

Males and females seem to court in the spring by playing a little hide-and-seek, flying back and forth from tree to tree. They both work to excavate the nest, a large hole in a dead tree.

Like other members of the woodpecker family, the downy eats a diet comprised mostly of insects. You will often see it moving energetically up and down tree trunks in search of food. Watch the way it uses its stiff tail feathers for support as it leans away from the tree. The downy is an acrobatic little bird, and can cling to very slender stems in search of the larvae that may elude the larger members of the woodpecker family.

Comfortable in open woodlands and deciduous forests, downy woodpeckers adapt happily to backyards and city parks. An offering of suet will easily attract this bird to your backyard.

Purple Martin

Purple martins and human beings share a unique bond. Over one million people in North America put up housing for purple martins, and in the

BELOW: Purple martin

eastern United States, purple martins have used these human-provided nest boxes for over a hundred years. The reasons people give for cultivating this unique relationship vary, but the answers share some common themes: Purple martins are great singers; they eat a lot of insects; they are loyal to their nesting location; they are acrobatic flyers; and they are actually friendly with humans. Whatever your reason for attracting martins, you are in good company!

The purple martin is a large swallow with a big head, thick neck, and broad chest. Its pointed wings and distinctive swallowtail make it easy to spot. The darkest of swallows, it has a deep, rich bluish-black color, and it's the only swallow with a dark belly. The adult male is the first to arrive in a territory, and he sings his *dawn song* to attract other martins to the area. Male martins have courtship flight patterns, often flying into their compartment and singing from the entrance to attract females. Older males may mate with females from other pairs, resulting in more than one clutch per season.

The female is a duller version of the male. She is also bluish-black and has a gray-brown belly. Juveniles have a similar appearance. Juveniles don't mature until the second year, so what appears to be a flock of females is probably a mix of females and subadult males. Martins are colonial nesters, which means that unrelated birds will share a common breeding site. A martin colony is just a group of unrelated birds attracted to a common breeding site.

Although purple martins choose to live together, competition for mates and nesting sites is stiff, and squabbles are inevitable. Female martins will sometimes steal nesting materials from other females, and older martins may steal food brought for fledglings.

After generations of adaptation, purple martins live almost exclusively in nesting boxes. It isn't easy to attract martins, but once you do, they will reward you by returning year after year. If you have a "colony" of purple martins—by definition, two or more breeding pairs in your nesting compartments—you are officially a "martin landlord."

Martins prefer open areas with plenty of room for taking off and landing. Unlike most birds, they actually have no problem with their nesting sites being near humans' homes.

Purple martins eat flying insects and can be very entertaining when out circling the sky for food. They are not attracted to feeders, although they can be conditioned to take live crickets or mealworms from your hand.

Cedar Waxwing

Seeing a cedar waxwing feasting on the berries in my dogwood tree on a cold November day is such a thrill. This extraordinarily beautiful bird, with its warm, multicolored tones, dapper black mask, and understated crest, often makes me think he deserves the award for best-dressed bird in the forest. Males and females have the same markings, and even juveniles wear the clearly defined mask. Even at a distance, his long, pointed wings and short, squared-off tail make him pretty easy to spot. But you might notice his call before you see him, as his high-pitched whistle is an easy one to learn.

Males and females carry out an affectionate courtship routine of passing berries, flower petals, or insects back and forth repeatedly. They check out nesting sites together, although the female makes the final decision and does all the nest-building. The cedar waxwing is not a very territorial bird, so you may note several nests within close proximity to one another. Nesting occurs a little later than it does with some birds, perhaps in keeping with the ripening fruit that these birds rely on for the bulk of their diet. They will also hunt insects throughout the spring and summer months, but they do feed their nestlings primarily a fruit-based diet. Cedar waxwings that find a batch of overripe berries can actually eat until intoxicated.

Cedar waxwings are common in forests and along streams, and anywhere fruit may be available. They are year-round or winter residents of

ABOVE: Cedar waxwing

most of the United States. They are a sociable bird, often traveling in flocks. The best way to attract them to your yard is to provide juniper trees and fruit-bearing trees and shrubs.

Dark-Eyed Junco

The dark-eyed junco is commonly known as the "snow bird" because of the likelihood of seeing it around your feeder in the winter. If you have a backyard feeder, chances are you have attracted dark-eyed juncos. One of the most common birds in North America, its migration is limited to the United States and Canada, so depending on where you live, the junco may very well be a regular customer. In many areas, they are full-time residents, but only come "to town" during the coldest months.

The dark-eyed junco is a medium-sized sparrow with a large round head. In spite of variations in color, it is hard to mistake this bird for another. They are those little gray birds on the ground below your feeder! Mostly dark gray or brownish with a white belly, you can make a positive identification by noting the pale pink bill and a distinctive flash of white on their tail feathers in flight. Males and females have the same markings, but variations abound. It's similar to the black phoebe in its coloring, but its attitude, habits, and posture are completely different.

The male junco establishes his nesting territory in the spring and awaits the arrival of potential mates. He is quite aggressive to other birds during the nesting season, and courts females with great displays of wing flapping.

BELOW: Dark-eyed junco

Once mated, the female builds her nest on the ground or on a sloping hill-side. Occasionally, she may choose to build in a tree or on an elevated horizontal surface, such as hanging planters or window ledges.

Dark-eyed juncos are birds of the ground. Look for them on the forest floor or hopping around below your feeder in search of small seeds or insects, often making little *chip* sounds as they eat. They are social birds outside of mating season, so you can usually see them in mixed flocks, eating side by side with other sparrows.

At the feeder, juncos prefer millet to sunflower seed. To attract them, lightly sprinkle millet on the ground below the feeder or on a platform feeder.

Tufted Titmouse

A common bird at the feeder, the tufted titmouse's gray appearance belies its perky personality. A good family bird, chances are the tufteds in your

BELOW: Tufted titmouse

yard are related to one another. Mom, Dad, and at least one offspring often stay together and feed throughout the winter. Look for the smaller juveniles at the feeder in the fall.

The tufted titmouse is a soft gray color above, with a white belly and rusty stains on the flank. It has a large head with a marked crest and a dark eye that gives it an intelligent, alert appearance. Despite its smaller size, its busy demeanor makes it seem dominant at the feeder, where it flits back and forth actively, snatching and hiding seeds or taking them to a nearby branch to crack them open. Like the chickadee it often hangs out with, it seems genuinely comfortable with humans around, and will often be the first one to investigate a change in the yard.

The female has the same markings as the male. Tufted titmice mate and stay together all season, usually having just one brood per season. The female chooses nesting holes left behind by woodpeckers or other birds. In my yard, they use the dog fur we leave out for them to line their nests, and have been known to pluck it directly from the back of a sleeping dog. If a nesting hole is not available, she may make use of a nesting box, drainage pipe, or gutter.

Common to deciduous forests and open woods, tufted titmice are comfortable in parks, backyards, and orchards. Year-round residents in the eastern half of the U.S., there are variations such as the oak titmouse, the juniper titmouse, and the black-crested titmouse found throughout the West.

Tufted titmice are omnivorous, eating insects in the summer but also readily eating nuts and seeds. At the feeder, this bird is partial to black-oil sunflower seeds, which they will hoard in stashes around the yard for later use.

Red-Tailed Hawk

While some smaller hawks may cruise by your feeder to see what's going on, it is more likely that you will spot the red-tailed hawk when you are out for a ride in the country. One of the most common hawks in the U.S., the red-tailed hawk can often be seen circling above open fields or posted atop a utility pole, watching the ground intently. They have a slow, heavy wing flap and an easily recognizable raspy scream.

The red-tailed hawk is a large hawk with a broad wingspan. It is brown all over, with a lighter-streaked belly. Its rather short tail is brick-red on top, but it's more likely you will see the underside, which is a pale-pinkish color. Females share the same markings but are larger than the males.

If you like high-drama courtship, this may be your bird. Males fly high in the air and then make precipitous drops. The male and female chase and swoop through the sky, sometimes clasping their talons together and plummeting toward the ground in big loops before releasing at the last moment. Pairs stay together until one of them dies.

Both members of the pair work on building a large messy nest at the tops of trees, utility poles, or other lofty perches, although they do sometimes reuse nests. The male usually does the hunting for the family during nesting, bringing food to both the female and the nestlings.

Birds of forests and open fields, these hawks breed throughout North America. Their diet consists primarily of small mammals, but they will also prey on larger birds. Farmers have nicknamed these birds "chicken hawks," undoubtedly because of their occasional poaching of the henhouse yard.

Rose-Breasted Grosbeak

The male rose-breasted grosbeak is a handsome, medium-sized songbird. Although he is an eye-catching bird, with his dapper black head, rosy chest, and white belly, chances are that you will hear him before you see him. His name refers to his distinctive large beak, which is thick and short and pale in color, and used for crushing seeds. The female is much harder to identify, looking very much like a plain, drab sparrow. Her pale bill may help to differentiate her from female finches or sparrows.

The rose-breasted grosbeak is known for its excellent singing, often referred to as sounding like a robin's song, sung by an opera singer. Males sing almost continuously during courtship, and males and females sing softly to one another while nesting. The male selects the nesting site, usually in a small tree or shrub, and the female builds the nest with his help. They incubate eggs and raise their young together. If the pair raises a second brood, the male will continue to care for the fledged brood while the female starts a new nest.

Rose-breasted grosbeaks make their homes in woodland areas and forest edges. Their diet includes insects, fruit, and seeds. They will occasionally stop by a feeder for seed, but they are infrequent guests, preferring the seclusion of the forest, particularly in mating season.

BELOW: Rose-breasted grosbeak

Common Crow

The common crow is anything but common. This remarkable bird has a big personality and a gregarious nature that makes him indisputably interesting. Almost always traveling in groups, this large, solid black bird is easy to spot, with its broad wings, long legs, and square tail. This bird is sometimes confused with the common raven, which is larger and has a heavier bill and a tapered tail. The crow is one of the rascals of the bird world, getting into mischief by foraging in garbage cans, stealing dog food, preying on nests and even smaller songbirds. The common crow is a clever bird that seems to enjoy problem-solving, and has even been known to fashion tools for reaching objects and chasing away other animals.

The male crow performs a preening and feather-fluffing display when courting a female. He bows and sings a short rattling song. Once accepted, the two birds will perch together and preen each other's feathers. Males and females work on the nest together and share in feeding the nestlings. Because crows often wait until their second year to mate, the previous year's offspring will also often help with raising the young.

Crows can be found, often in extremely large numbers, across most of the United States. They like open farmland, light forest, riverbanks, and pastures, but have become very comfortable in city parks, at landfills,

and on golf courses. Because they tend to travel in large groups, their noise and their mess sometimes make them unwelcome visitors.

Crows seemingly will eat just about anything, from garbage and earthworms to fruit, mice, and seeds. Although they are not steady customers, you

ABOVE: Common crow

will sometimes see them at the backyard feeder, where they will help themselves to whatever is available and intimidate your other regulars.

Mourning Dove

This is one of the most common birds in North America, and in most areas is considered a game bird. I like to count these gentle little birds as some of my favorite backyard visitors. Plump-bodied, with a distinctively small head and a sweet cooing call, how can anyone fail to enjoy them? Mourning doves are buff-colored overall, with black-and-white accents. Males and females share the same appearance and can often be seen perching together on a branch or wire.

Males and females work together on the nest, with the male bringing materials and the female doing the building. Incubation and feeding are also shared. Parents regurgitate a substance called pigeon milk to feed the newborns, gradually supplementing it with seeds. Parents continue to feed the fledglings for another two weeks or so after fledging. The mourning dove is an attentive parent, and nests are rarely left alone. Clutch sizes usually consist of only two eggs, and pairs will often have several broods in a season, sometimes using the same nest. Nesting sites are chosen together, and are often within easy proximity to humans. Nests can be found in low branches or in gutters or hanging plants. Provide a nesting basket in the

BELOW: Mourning dove

crotch of a tree to attract nesting pairs. Not highly territorial, sometimes pairs will nest in the same tree.

Mourning doves live in open country and the edges of woodlands, and often roost in groups during the winter. They are ground foragers that eat a diet composed almost entirely of seeds. They will often stay at the feeder for long periods, picking up seeds to fill their crops, and then flying off to perch somewhere and digest their seeds. At the feeder, you can scatter millet on the ground to attract them.

European Starling

Not native to America, the European starling was actually imported here in the 1800s by fans of Shakespeare who wanted birds that were represented in his plays. These birds have taken to North America quite well, and are now actually one of the most common songbirds, making their home comfortably across the entire continent. Often resented for their abundance and their noisy nature, a large flock of starlings can make a huge and noisy mess of urban streets and city parks.

Although a similar size and color to grackles and other blackbirds, European starlings can be easily identified by their shape. They have a shorter tail and a longer, more-slender yellow bill that distinguishes it from others of its type. Their wings are short and pointed, creating the appearance of a small, four-pointed star at the tip. They look black from a distance, but are actually an iridescent purple and green in the summer. They may appear brown in the winter.

European starlings nest in cavities. The male actually builds a nest to attract a female, singing as he works, and often decorating it with flowers and greenery. Once the female accepts him, she often

RIGHT: European starling

redecorates, discarding some of the materials he has added. Male and female incubate and feed together, and will often reuse the nest for another brood. Starlings are aggressive and will colonize purple martin colonies, driving away the native birds and marauding existing nests.

European starlings prefer a diet comprised mostly of insects, but will eat berries, grains, and even garbage. They are common in cities and near human habitats.

Indigo Bunting

The romance of the indigo bunting lies in his magical color as well as his habit of migrating at night with guidance from the stars. The indigo bunting learns its orientation of the stars as a young bird, and stories tell of captive birds actually becoming disoriented when they are unable to see the night sky. The indigo bunting's shock of electric blue is always a thrill to see in the spring. In this species, it's all about the male; his spectacular spring plumage is sure to catch the eye of an available female. Females take a more-subtle approach, with an understated all-over brown appearance. Off-season males have a similar appearance to females, with juveniles taking after Mom for the first year.

BELOW: Indigo bunting

Female indigo buntings do all of the nest building and incubation, although males may help feed fledglings while the female starts on a new nest. Nests are generally built in shrubs or low branches. Indigo buntings favor weedy grasslands and open woodlands. Their diet consists primarily of insects, although they supplement their winter diet with seeds and fruit.

Brown-Headed Cowbird

The brown-headed cowbird is a stocky blackbird with a thick head and a short tail. It takes its name from its preference for pastures and open fields, where it eats grain seed and snatches grasshoppers and other insects kicked up by herds of cattle or bison. The male has a glossy black body with a rich brown head. Females are a simpler all-over brown color.

Not the most popular bird on the block, the brown-headed cowbird has an unusual breeding plan. Females lay as many as three dozen eggs every season, depositing a single egg at a time into the nests of other species, leaving them for others to hatch and feed. Some species are wise to this and will throw out the eggs, but most species raise the nestlings as their

BELOW: Brown-headed cowbird

own. Cowbird nestlings are aggressive, too, maturing faster than their host siblings, and often killing them or rolling eggs out of the nest.

Brown-headed cowbirds like grasslands and the edges of woodlands. They are not likely to be found in forests. Once primarily a western bird, brown-headed cowbirds can now be found across all of North America.

Eastern Meadowlark

Not a lark at all, but rather a member of the blackbird family, this charming little flutist is a familiar sound on farms and meadows with his cheerful "Spring is here!" song. The eastern meadowlark is a medium-sized songbird with a short tail and an alert, optimistic posture. It uses its long, pointed bill for snatching grasshoppers and other field insects. The centerpiece of his appearance is a bright yellow chest with a distinctive black V. He has a streaky brown-and-black back and strong eye stripes. Males and females have a similar appearance, and juveniles are more muted overall. The eastern meadowlark is similar in appearance to his cousin, the western meadowlark. Location will give you a good idea of which one you are looking at, although their ranges overlap somewhat.

Males usually arrive earlier in the season to stake out territory before the females arrive. During mating season, male meadowlarks will choose a fence post or a high rail from which to sing his song. Males often have two mates per season, claiming a territory of six to seven acres for them. The female builds her nest alone, weaving it right into the landscape of a grassy area. It generally has a side entrance and an arched or roofed entry. She does her best to hide her nest from predators. Unfortunately, one of the

BELOW: Eastern meadowlark

most common predators for meadowlarks is the deadly mowing machine. Nesting usually occurs in July and August, and fields that are hayed before nesting has been completed kill many birds every year.

Once considered a common bird in rural areas, the eastern meadowlark's numbers have been declining in recent years, probably due to habitat loss. Look for them in open fields and meadows, or near golf courses.

White-Breasted Nuthatch

Who's that bird hopping down the tree headfirst? Chances are pretty good that it's a nuthatch, one of the few birds comfortably capable of maneuvering up, down, and sideways along a tree trunk. It often looks up and around when descending a tree, giving it the appearance of a jaunty, debonair attitude, although it's more likely the bird is checking for predators.

The white-breasted nuthatch is not only acrobatic; he is also an elegant little bird, always looking like he is dressed for dinner. The wide white

White-breasted nuthatch

breast looks rather like a tuxedo shirt, accented as it is by his black cap and blue-gray body. Males and females stay together throughout the year, foraging with nuthatches and titmice in the relative safety of their company.

White-breasted nuthatches are secondary cavity-dwellers. That means that they look for old woodpecker holes or other natural cavities in which to build their nests, rather than excavating their own. They may even use your bluebird nesting boxes. The female builds the nest herself, filling the cavity with dirt and bark and then lining it with a softer layer of feathers, grass, and fur. The male feeds the female while she is incubating the eggs, and both parents work together to feed the nestlings. Although white-breasted nuthatches only have one clutch per season, it is a relatively large one, generally containing five to nine eggs. Babies fledge a little later than many songbirds, as late as twenty-six days after hatching, compared to the more-typical fifteen to twenty days of the eastern bluebird.

White-breasted nuthatches prefer the habitat of a good, old-fashioned deciduous forest, and eat wood-boring insects, caterpillars, and larvae, as well as seeds and acorns. At the feeder, they will come for sunflower seeds and suet.

White-Throated Sparrow

The white-throated sparrow is well known for his song, a wistful tune variously interpreted as *Poor Sam Peabody-Peabody, Peabody* and *Oh Sweet Canada, Canada, Canada.* Because white-throated sparrows generally head north to Canada to breed, I choose to go with the latter; I like to think of it as a melancholy reminiscence of mating season and past love.

The white-throated sparrow's field markings give you a lot to look at, from its head's strong eye stripe, its colored crown, yellow lores, and white throat, to the distinctive white wing bars. There are two variations in the head markings: a strong black-and-white striped version, and a softer buff-

ABOVE AND RIGHT: White-throated sparrow

and-brown stripe combination. Males and females have the same markings. Females prefer tan-striped males and males prefer white-striped females, but according to the Cornell Lab of Ornithology, white-throated sparrows have even been known to mate with the totally unrelated dark-eyed junco.

White-throated sparrows prefer a habitat of dense, shrubby vegetation and nest on or near the ground; they will readily make use of a brush pile for cover or nesting. They are ground foragers, dining companionably in flocks with dark-eyed juncos and other ground birds. Their natural forage ranges from seeds and fruit to summer insects. At the feeder, they enjoy both sunflower seeds and millet.

House sparrows

> **"**'Hear! Hear!' screamed the jay from a neighboring tree, where I had heard a tittering for some time. 'Winter has a concentrated and nutty kernel, if you know where to look for it.'**"**
> **—HENRY DAVID THOREAU**

4

Backyard Bird Feeding

THE FUN is about to begin! The very best way to start on your journey toward understanding birds is to have the birds come to you. Now is the time when learning what they like to eat, how they travel, and where they live provides huge dividends for you.

How will the birds hear the news about your feeder? It is probable that your first visitors will be one of your small, local, year-round birds. Chickadees, nuthatches, and titmice are always on the lookout for a new food source, and will probably find you first. It may take a few weeks for more birds to find your feeders. After that, you may see the floodgates open! Other birds will notice the activity and be drawn to check it out.

If you notice that interesting birds are stopping by and not staying, take time to learn about what they like to eat. Some of our favorites, like cardinals and juncos, may prefer to eat on the ground. Finches and pine siskins are on the lookout for weed seed. Woodpeckers don't have a lot of interest in seeds, but they will definitely become regular visitors to your suet feeder. Once your feeder program is in full swing, you won't have any problems attracting customers.

LEFT: A familiar face at feeders across North America, the chickadee's affable nature makes him a welcome part of the mixed flock.

Bird Food Basics

Bringing birds to your yard is really the best way to get to know them. Feeding birds is a big industry, so you may feel a little overwhelmed the first time you visit your local feed store, nursery, or home center. Along with a wide array of decorative and specialty feeders, there are a lot of seeds, nuts, and commercial birdseed mixes on the market. All those choices! If you are a natural-born shopper, it may look like paradise, but it's not really necessary to spend a lot of money to start your feeding program.

Food Types

With all of those seeds, how do you choose? Here are the basics:

Black-Oil Sunflower Seeds

These seeds are the "meat and potatoes" of your bird feeder. A little different than the traditional striped sunflower seed that might first come to mind, the black-oil sunflower seed has a higher fat content and a bigger meat-to-shell ratio that makes it very attractive to birds. Jays, cardinals, woodpeckers, and grosbeaks all love them, but their softer shell also makes them appealing to small birds like chickadees, nuthatches, and titmice. The good news is that these seeds won't sprout beneath the feeder. The bad news is that they are so popular, there will be lots of empty shells to clean up!

Hulled Sunflower Seeds

These seeds are commonly packaged as sunflower hearts, pieces, or chips. Everyone at your feeder will love them, but these seeds are an especially thoughtful choice for redpolls, Carolina wrens, and goldfinches that may have a hard time breaking into shells. Because they are readily eaten and have low waste, they are great for places where you don't want to deal with the mess left by discarded hulls. More expensive than seeds in the shell, you will find that a little actually goes a long way. Don't put out too many at a time, though; they need to be protected from the elements and will rot quickly if they get wet.

White Proso Millet

Sparrows, juncos, mourning doves, and towhees are just some of the birds that enjoy this seed. It is best scattered on the ground, where these birds naturally forage. A common component of birdseed blends, you may

Starting Your Feeding Program

Not sure where to start? Your first shopping list should include:

- one simple tube feeder
- one bag of black-oil sunflower seeds
- one storage container with a tight lid
- one suet cage feeder and a cake of suet

Let the fun begin!

LEFT: Dark-eyed juncos are ground feeders that enjoy the millet often cast aside by other feeding songbirds.

find that a lot of the regulars that come to your feeder push it aside in favor of the sunflower seeds, so if you want to attract birds that prefer millet, buy it separately and use it specifically for them. A word of warning: cowbirds, blackbirds, and other annoying birds like house sparrows also like it, so if you don't want to attract them, stick to sunflower seeds. Also, millet can sprout beneath the feeder, so if this is a concern, you may want to sterilize the seed before using it. To do this, place one pound of seed in a brown paper bag and cook on high in the microwave for five minutes.

Safflower Seeds

Safflower seeds are a favorite of cardinals, and some other species eat them as well. The main reason to use safflower, though, is because of what *doesn't* like it. Feeding safflower may discourage squirrels and a few nuisance birds, like grackles and starlings. This is not a sure thing, however; I have friends who say their squirrels got over their aversion and now go after the safflower feeders as well!

Nyjer Seed

Also known as thistle seed, this food is essential for attracting finches and pine siskins, especially during migration. These tiny seeds are heat-sterilized so you don't have to worry about sprouting weeds underneath your feeder. They blow away easily and are kind of pricey, so tube feeders with tiny openings are made especially for this type of food. Nyjer seed needs to be fresh, so place it where it can stay dry and keep an eye on it for signs of mold. Give the feeder a quick shake once in a while. If the seed is clumping, throw it out and start fresh. Nyjer seed can also be offered in "thistle socks." These fine nylon sacks are relatively inexpensive, allow air to circulate around the seed, and birds like to perch on them.

Peanuts

Who doesn't love peanuts? Peanuts are gaining in popularity as a bird food, and a number of species love them. There are even feeders designed especially for dispensing shelled and chopped peanuts. Jays enjoy whole peanuts in the shell, as long as they can get to them before your squirrels do! Peanuts are not as stable as sunflower seeds, though, so you have to make sure they stay fresh. Immediately throw out any that are darkening, as this is a sign that they are going rancid. Also watch for signs of mold.

Corn

Dried corn, either whole-kernel or cracked, is useful if you are trying to attract wild fowl like turkeys, geese, quail, or ring-necked pheasant to your backyard. Blackbirds, cowbirds, and sparrows will also eat corn, but so will deer, raccoons, and fox.

Commercial Seed Mixes

You know how your kids sort through the trail mix to get at the M&Ms? Birds are kind of like that when it comes to seed mixes. There are a lot of blends on the market, and good-quality mixes can attract finches, sparrows, and songbirds. Blends can be fine, but skip the bargain-basement blends. They contain a lot of fillers that birds just don't care about, like corn, oats, milo, and other cereal fillers. Look for sunflower-rich blends to make your birds happy, and don't be surprised if certain birds pick through the feeder looking for the pieces they want.

Suet

After black-oil sunflower seeds, suet is the most important offering you can make in your backyard. It provides a vital high-energy fat food source that is especially useful for birds, not only in the winter when other sources are scarce, but throughout the year. Suet attracts woodpeckers, nuthatches, chickadees, and blue jays; and wrens, creepers, and even cardinals might stop by for a quick snack. Pure suet should only be put out when the temperatures are cool enough to keep it from melting or going bad. Not only is melting suet messy in your yard, but the greasy suet can damage belly feathers. If an adult brings that gooey mess back to the nest, it can damage eggs.

During the warmer months, there are a number of warm-weather "no-melt" alternatives that contain corn, seed, and other ingredients. You can buy commercial suet cakes or make your own.

BELOW: The eastern bluebird is not tempted by your seed feeder but can be lured in with an offering of mealworms.

Mealworms

Mealworms are actually the larvae stage of a small beetle and are "bird candy" to a number of species. Some insect-eating birds that aren't attracted to your traditional feeders may be willing to stop by for a quick meal of fresh, live mealworms. I use them to bring in eastern bluebirds, although once the word has spread, everyone may want to stop by! During nesting season, even seed-eating birds will appreciate this high-energy food source. They are expensive, though, so keep them for those birds that you really want to have them, and put out only a few at a time. If you don't like the idea of dealing with live mealworms, there are freeze-dried or roasted versions. Birds don't find them very tempting because they are attracted to the wriggling of the live ones. Some people try to make them more appealing by reconstituting them in olive oil or water. I don't think this method is particularly effective. I do use dried mealworms in bird food recipes or as a topping to a suet recipe.

For Birds with a Sweet Tooth

Nectar

To attract hummingbirds to your yard, you either need a garden full of flowers or an offering of nectar. Occasionally, orioles and a few other species may stop in for a sip. Nectar is a basic four-to-one blend of sugar and water which you can easily make yourself. *Never* use molasses, honey, or artificial sweeteners. Don't use commercial blends, either, as they usually contain red dye that is not necessary and may actually be harmful.

Fruit

A large part of a bird's diet is fruit, so many of them appreciate seeing a piece of fresh fruit at the feeder, especially in the off-season. Orioles love oranges, but I find that apples are more popular with other birds. I have seen bluebirds, catbirds, mockingbirds, robins, and woodpeckers checking out my fruit.

Cut a piece of fruit in half, string it on a wire hanger, and hang it in a tree. I give my apples a quick spray of lemon juice to keep them from browning. That's not for the birds, but for the sake of my photographs. I don't think the birds care whether the apples are brown or not.

In the winter, dried fruit is appealing. Almost any dried fruit, from raisins to apples and apricots, is a welcome treat. I dry my own "bird apples" from my friend's wormy apple tree. Most birds can't handle big pieces, so chop it up before serving. I find it easy to use a pair of scissors to cut up tiny pieces. Consider saving all of your overripe or past-their-prime berries in the freezer for a fresh wintertime snack for your birds. (Don't save anything with mold on it.) I even know people who cut whole stems of native bird berries and put them in their freezer for winter use.

Jelly

Jelly? Some birds have a serious sweet tooth and will be thrilled to find a small dollop of jelly left out for them. Don't do this too often, and only put out a little at a time—a big jar of inexpensive grape jelly should last a whole season. A tablespoon in a shallow dish is just fine. If you put out more, you may also attract ants and wasps.

Food Storage

Whatever kind of food you choose, it is important to keep it dry, fresh, and safe from pests.

Seeds

Keep your seed fresh to make sure that you are offering your birds the healthiest possible product. Seeds and nuts that develop mold or go rancid can be fatal to the large population of birds that count on you for food. The large economy-sized bags of sunflower seed might have an appealing price tag, but if you do not have a big feeding program, don't buy them. In warm weather, bring home only what you think you will feed in a couple of weeks. In the winter, a month's supply should be fine.

Store your seed in a cool location. Sunflower seed in the shell is pretty stable, but cracked grains and seeds are susceptible to faster deterioration.

Proper purchase and storage is not only good for the birds, but it will also help you avoid the problem of larval outbreaks—the last thing you want in your kitchen or pantry! Birds wouldn't mind the extra protein, but you may find it distasteful to deal with. I have a friend who adds whole bay leaves to her seed container. Though I can't vouch for it, she swears by their ability to deter insects.

A stray bag of seeds is like a magnet to squirrels and mice, so keep all your bird-feeding supplies in tightly closed containers. Plastic is okay indoors; we keep a large plastic snap-tight box in our kitchen for foods we use every day. If your family receives those big tins of popcorn for Christmas, reuse them for storing birdseed. For larger quantities that you plan to store outside, in your garage or your shed, use a metal can with a tight lid. If you think you might have visiting raccoons, add a strong bungee cord.

Nectar

Make one batch of nectar at a time and refrigerate any leftover nectar between feeder fillings. Keep your refrigerated nectar for a week to ten days.

Fruits and Nuts

Keep dried fruits and nuts in airtight bags and look them over for signs of spoilage before using. Most fruits and nuts will keep for six months or so, but even birds don't care for really old product, so discard anything that gets too old.

Suet

Commercial suet cakes are pretty stable, so they just go into our plastic box until we are ready to use them. Homemade suet and fresh beef suet should be stored in the freezer.

Types of Bird Feeders

On your first shopping excursion for bird feeders, you may find the array of choices bewildering. Because birds have preferences based on food type and foraging style, feeders are designed to accommodate their needs. And you have some decisions to make before shopping, too. How big should your feeder be? Are you going to start with a small amount of nyjer seed or a large supply of sunflower? How will you hang it? Is it easy to clean and refill? Is it durable?

Keeping all those factors in mind, it's time to go shopping. Though there are many variations, seed feeders fall into three basic categories:

Seed Feeders

Hopper Feeder

Hopper feeders come in a variety of shapes and sizes, but all have the same common design—a large central storage container that dispenses seed onto a tray at the bottom of the feeder. Hopper feeders often have a roof for keeping seed dry, and perches on both sides to accommodate several birds at once. Many are designed to hold a large amount of seed. They are easy to fill and clean.

Tube Feeder

Tube feeders have feeding ports and perches that allow small- and medium-sized birds to perch and eat in comfort without getting shooed away by bigger birds. Some tube feeders have interchangeable ports specif-

ABOVE: Hopper feeders come in many styles. This stylish copper feeder has caught the attention of a Carolina chickadee.

ically for sunflower seeds or nyjer seeds, but most tube feeders are designed specifically for one or the other. Look for variations that include a perching ring at the base to allow bigger birds to feed, or a small roof or weather dome to allow birds to duck out of bad weather.

Platform Feeder

Also called table or tray feeders, these are simple flat surfaces, often with a roof for keeping out the elements. Platform feeders are generally mounted on a post, but positioned lower to the ground, they can be attractive to ground-foraging birds. Some of the more-sophisticated platform feeders have screen or perforated bottoms that help seed to stay dry. Besides seed, these flat feeders are a good place to offer fruit or bread scraps.

RIGHT: Sort of a birdie gumball machine, the tube feeder dispenses seeds neatly and provides an individual perch for each port. This tufted titmouse is happily taking his turn.

Specialty Feeders

In addition to the standard feeders, there are several specialty feeders, designed to deliver goodies to certain types of birds. These are some of the most typical specialty feeders:

Peanut Feeder

Designed specifically for dispensing peanuts, these feeders generally have a wire-mesh cage that birds can cling to while they enjoy their peanut pieces. Other peanut feeders hold whole peanuts. Peanuts have become a much more popular food in recent years, as peanut producers have realized that bird feeders are a great market for broken and non-human-grade nuts.

Fruit Feeder

It is really not necessary to have a feeder specifically designed to offer fruit. A cut-up orange or apple put out on a picnic table or platform feeder will work perfectly well. Be careful, though; an overlooked piece of fruit on a tray feeder can rot and introduce bacteria and mold to the surface of the feeder. Hanging feeders that are designed specifically for fruit keep the offering clean and allow you to place fruit where birds can perch to enjoy it. They also hold the fruit securely so a larger bird or mammal can't run off with it.

Hummingbird Feeder

Hummingbird nectar feeders come in a variety of designs, but two basic shapes. Basin-style feeders are shallow covered trays, often with perches. They can attract ants, so some have built-in ant moats. Inverted bottle feeders are long and narrow and generally require the bird to hover to sip the nectar. Choose a feeder that is easy to take down and clean. Hummingbird feeders should be thoroughly cleaned every time they are refilled. Nectar needs to be fresh, so choose a small-size feeder until you know you have hummers in the neighborhood. Hummingbirds are fiercely territorial, so you may find that you will be more successful with a series of small feeders than you are with one large one. Our flower garden has several small bottle feeders positioned at the same height as the flowers.

Mealworm Feeder

Feeders designed specifically for offering mealworms range from simple shallow cups with straight sides to boxes that allow bluebirds to fly in and out to get their treats.

Finch Feeder

Small birds that eat weed seed love nyjer seed, a tiny lightweight seed that has to be fed in a tube to keep it from blowing away. There are several finch feeder sizes available. Nyjer seed is relatively expensive and molds easily in wet weather. If you are new to feeding nyjer, start with a smaller tube to save seed, and then add an additional tube if you find that this feeder is becoming popular. You can also use a finch sock, which is a simple nylon mesh bag filled with nyjer seed. Because finches tend to congregate in groups, you may soon notice overcrowding at this feeder. Consider positioning another feeder away from the first one to break up the crowd.

LEFT: This attractive cedar tray feeder blends nicely into the environment, and provides birds with a welcome feeding area, away from the hustle and bustle of the tube feeders.

CENTER: An unusual moment at the finch feeder; once word gets out that you offer nyjer, these sociable birds will fill every perch.

RIGHT: This bottle feeder only allows one bird at a time to feed—not a bad idea when feeding fiercely territorial hummingbirds!

Suet Feeder

There are a lot of different ways to offer suet. The most basic is the inexpensive wire cage that holds blocks of commercial suet cakes. You can also use suet balls or mesh bags. We like to use a log feeder. This is a simple log with holes drilled in it for holding suet or peanut butter.

Where to Put Your Feeders

Put up feeders in places that allow you to enjoy the birds that come to visit. Find window vantage points and then look around for appropriate settings near those windows. Keep the feeders in a convenient location. If you put them too far from the house, you won't be as anxious to fill and tend them, particularly when the snow piles up. You need to be able to take feeders down in order to clean them, so it's best to use a simple hanging system. Our largest feeder station is under a stand of trees not too far from the quiet side of the house, away from the coming and going of human traffic. There are branches and shrubs for visiting birds. The ground below is mulch and easy to rake up. A large window is nearby, so watching and taking pictures is convenient. Birds can come and go and make as big a mess as they want—it's *their* area.

We also have a tube feeder and a nectar feeder near the patio, and suet logs in the nearby trees—a more-intimate location for having coffee in the morning with our birdy friends. Our grandson has a window feeder for

hulled sunflower seeds. Hulled seeds are nice in window feeders or for apartment or condo dwellers because there are no hulls to leave behind, and no unwanted seeds for birds to pick through and toss aside.

If you have a pristine lawn, give careful consideration to the ground directly underneath your feeder. The seed shells that accumulate under your feeder can kill the grass, suet can stain decks, and bird droppings can cause an

LEFT: We love the look of this handmade suet log almost as much as our birds love using it.

unsightly mess. The area beneath feeders should not be in a direct path for humans and should be easy to rake up. If you do want a feeder in a place that might harm the lawn, there are bird-feed catchers that can be mounted beneath your feeder to contain the mess.

After the human considerations are met, think about what your birds want when selecting your feeding location. Their needs are simple, but accommodating them will reward you with more birds and better viewing. Look for a spot that has the following criteria:

- **Sunny:** A sunny location will be pleasant for winter birds and offer you good viewing. Find a protected spot away from strong breezes.
- **Sheltered:** Look for places in your yard within easy reach of protective cover. Birds like to have a comfortable place to wait in a holding pattern for their turn at the feeder, so shrubs or a nearby pine tree is just the ticket.
- **Safe:** Cats and other predators can jump as high as 10 feet to snatch at feeding birds, so make sure they don't have an easy jumping-off point to feeders.

Maintaining Feeders

Because your feeder is a very public and social location for an array of bird species, it is important to maintain a clean and healthful environment. There are several reasons for this. Moldy or rancid seeds can make birds ill.

Aspergillosis is a mold that grows on damp feed. Birds inhale mold spores which cause an infection in the lungs, resulting in bronchitis and pneumonia. Debris underneath the feeder can also grow these molds, so ground cleanup is important as well.

Just like humans congregating at schools, day-care centers, or other public places, visiting birds carrying illness can spread salmonellosis and other avian diseases. If you have a robust feeding program, help reduce the chance of spreading illness by adding feeders to prevent overcrowding.

Finally, birds prefer clean, fresh food just like we do. Keeping a clean feeder with fresh food is the best way to attract and maintain bird visitors year-round. And they are generally pretty savvy too. If you notice birds stopping by, taking a look, and then leaving without eating any food, you may want to investigate the freshness of your food.

Maintaining Hummingbird Feeders

Attracting hummingbirds is a thrill that never gets old. If you have never fed hummers before, it may take a little while for them to find your feeder, but once established, they will be remarkably loyal to you. Make sure you earn their trust by taking good care of them. Make your own nectar rather than buying the commercial mixes. The red dye in those blends is completely unnecessary and may actually be bad for the birds. A simple four-to-one ratio of sugar and water is all that is needed. See the chapter on recipes to get details for making it. Keep leftover nectar in the refrigerator.

Choose a shady location out of direct sunlight to keep your nectar from spoiling quickly. Hummingbird feeders should be refilled and cleaned often during their season. Take them down every week and dump out any leftover nectar before thoroughly washing the feeder with soap and water. Use a brush for narrow openings, or choose the saucer-shaped feeders that are easy to clean. You can also throw it in the dishwasher.

Cleaning Seed Feeders

Every time you fill the feeder, dump the leftovers and look the feeder over for damage or evidence of dampness. Squirrels and woodpeckers can chew or peck through plastic, and a knocked-down feeder can have a cracked tube. Replace damaged feeders. Clean off bird dropping or other debris. I wipe down each perch with a solution of bleach and water. Don't hesitate to do a more-thorough cleaning if you think it's needed.

To do a thorough cleaning of your feeder, take it down and dump all the contents. I use an outdoor hose and nozzle to knock off all dirt and debris both inside and out. Then I make a solution of soapy water and bleach and scrub all the surfaces. A ratio of one cup of bleach to nine cups of water is a safe bleach solution. Use a soft sponge for outside surfaces and bottle-style brushes for inside tubes. Rinse thoroughly with fresh water and dry off with a towel. Set feeders out in the sun to dry thoroughly. Don't refill feeders until they are completely dry.

I do this kind of "deep-cleaning" of my seed feeders once a season, or four times a year. You may want to clean them all at once, but if you have a lot of feeders, you can do a few at a time.

Feeding Programs by Season

The life cycle and migration patterns of birds create different needs for each season. Early spring arrivals have different requests than your year-round birds. Weed-seed eaters drop by later and want weed seeds. Winter birds are still looking for some of their summer and fall favorites. Watching and learning about the needs and lifestyle of your birds will help you to learn how to supplement their diets appropriately, and help you to attract visitors that you would like to see.

Follow this plan for a healthy bird-feeding experience year-round.

Autumn Feeding

If you are just getting started, this is the season to begin. Migrating birds appreciate the convenience of a quick, high-fat meal as they travel south, and year-round residents are happy to find that you will be a source of their winter's meals. Start with basics like black-oil sunflower seeds, peanuts, and suet cakes.

If you already have a feeding program, now is a great time to do a review. After a long summer of relative inactivity, feeding season is about to start in earnest. Check all of your feeders and replace any broken ones. Add any other feeders that you want to use this season.

Offer hummingbird nectar throughout the fall for migrating birds. As flower supplies dwindle, nectar feeders become more attractive, so I add extra feeders starting in August. In Virginia, my feeders stay up until November to make sure I have a meal available to migrating birds, but when to take them down varies regionally. When you take yours down, clean and sterilize it before storing it for the following season.

I put more seed feeders up in anticipation of heavier traffic. In September, I deep-clean all feeders that have been up throughout the summer, and I re-clean feeders that have been in storage. If I have had suet feeders up, I do a thorough soak in a grease-busting detergent. The holes in my log feeders get a spray and scrub before refilling and hanging them. Wooden platform feeders need a soapy scrub, too. Put your wooden feeders in the sun to dry thoroughly before you put them back out.

Take a look at your feed supply and discard any seeds that are looking old. Sunflower seeds in the shell are good for a year. Nyjer seeds are good for only six months or so. Suet and suet cakes should be stored in the freezer and will keep for up to a year. I always freshen up my supply of dried fruits and shelled nuts at this time.

Winter Feeding

Now is the time when your birds are counting on you to come through. Maintain a rich assortment of seeds, white proso millet, and suet at all times. High-calorie treats are very welcome at this time of year, so think fat! Suet, peanut butter, and lard mixtures will all be popular. Fruits, both fresh and dried, are also much appreciated. I also offer more nuts in the winter.

Winter is an especially rewarding time at the feeder, because birds tend

to congregate in flocks to find food. Once the migrants have moved on, your visitors will probably settle into a familiar group of mixed birds. In cold weather, and without much natural food available, they seem to linger longer at the feeder. This gives you the opportunity to study some of your favorites a little more closely. The

LEFT: This gentle mourning dove perches patiently, waiting for its chance at the feeder.

ABOVE: Sudden winter storms put extra stress on birds who especially appreciate a chance to stock up at your feeder. These mourning doves and cardinals are waiting for their chance at the seed.

chickadees, nuthatches, and titmice come together, but they are not averse to allowing other visitors to join them at the feeder. Woodpeckers and juncos may also be regular visitors. Depending on your region, you will have other regulars. This is a great time to learn their names and study their habits.

Keeping your feeders clean is really important now. Make sure that there is nothing in your feeding program that can cause the spread of illness. Dump seed that is old, and if you feed fruit or bready foods that can mold, make sure you clean them up before they make someone sick. I usually do my deep-clean of feeders in late December. Because I am filling feeders more often, I clean one feeder at a time, so there is always a food source in place. Now is a good time to rake up below your feeders before the snow gets too deep.

If you do notice a sick bird, it may be a good idea to take in feeders for a little while. Although you might really want to help the sick bird, it doesn't make sense to have it in there feeding with all the others and contaminating the feeding area.

ABOVE: The blue jay is a year-round resident, and a familiar face at the feeder in every season.

Springtime Feeding

In spite of the burgeoning food supply in the backyard and forest, spring is no time to stop providing food. Your year-round residents are still stopping by for their sunflower seeds, cracked corn, millet, and suet. And guests traveling through will be thrilled to find a meal waiting, providing you with the pleasure of their company for a few days—or even a few weeks—before moving on to their next location.

Spring is an exciting time for seeing migrants. Depending on where you live, you should see flocks of finches, warblers, and sparrows sometime in April through early May. Take time to watch for their arrival and enjoy them while they are there, because who knows how long they are going to stay! Because they migrate at night, you may wake up one morning to find an empty yard and an abandoned feeder. This time of year, I have a friend who often calls in to work with "warbler flu." She can't bear to miss the passing flocks!

Finches are a delight at the feeder, especially in the spring when they are decked out in their best colors. Members of the finch family prefer tiny nyjer seeds, so a tube or feeder sock is the best way to attract them. It is important to keep this tiny seed fresh and dry, so check your feeder often and discard what hasn't been eaten within a month.

Even though you might not be able to identify which one is which right away, you will know when the warblers arrive. If the yard suddenly has a frantic, busy air that it didn't seem to have before, the warblers may have arrived. And those flashes of color that you didn't see before are a telltale sign. Warblers are not big feeder customers, since their primary diet consists of insects, but they can be attracted by mealworms and suet.

Wicked Weather and Late-Season Storms

The problem with spring is that the weather is often unpredictable. Birds that made the decision to move north sometimes find themselves caught in foul weather or a late-season storm that wreak havoc on their food supply. Be prepared to supplement their needs with a special variety of foods for times like this.

If a storm hits, reach for these ingredients from your pantry:

- Peanut butter spread on tree trunks or pinecones (a great high-calorie snack)
- A crumbly mix of cornmeal, dried fruit, and peanut butter for bluebirds and thrushes
- Crushed nuts, such as walnuts, pecans, or peanuts, for small birds like wrens and catbirds
- Larger nuts for nuthatches and jays
- Oranges, bananas, and apples for orioles and thrashers
- Grape jelly for orioles
- In a pinch, a tray of moistened high-protein dry dog food will attract bluebirds and starlings.

If you are concerned about a late-season storm, have suet and mealworms available. Even if they are not a part of your regular feed, having

Winter Vacation?

If you are a winter bird yourself and plan to take an extended vacation to other climes, keep in mind the responsibility you have to the birds that depend on you. If you are in a neighborhood with other bird-feeding friends, or in an area that has some natural resources for your birds, a three-week vacation is not a big deal. But if you are in a cold climate, are the only local bird restaurant, or a winter storm comes up while you are away, a sudden lack of food can be dangerous for the bird residents that have come to depend on this food source.

them on hand can keep your feathered friends from starving or freezing to death.

Summertime Feeders

Good weather is bringing out the insects, fruit and berries are ripening, and the nesting season is on. The birds that you have been accustomed to seeing have paired up and gone off to set up housekeeping together. If you have made your yard into an attractive habitat, you may still have lots of birds around, but summer is not primetime feeder season, so I reduce the number of feeders I keep out, and stick to the basics. Sunflowers and millet are my primary menu items, and I generally don't have a lot of takers. Still, there are a few regulars who like to stop by and grab a quick seed or two for the road.

Suet can be messy in the summer, especially in the heat of Virginia. Melting suet stains decks and sidewalks. Birds do appreciate the opportunity to stop by for a quick high-fat snack, so I switch to no-melt suet cakes that resist melting and still appeal to most of my suet customers.

The natural food supply is abundant right now, and nesting birds have to make the most of that. There are several mouths to feed back at the nest, and, for most baby birds, insects are their food of choice. Even birds that normally eat seeds will be out in search of insects for their nestlings. Because it takes so much energy to raise their young, adults may appreciate the chance to stop by for a few seeds for themselves, and may reward you by bringing out their family to visit you and your feeder on one of their early outings. Fledglings enjoy soft treats such as suet and peanut butter while still getting the hang of dealing with seeds and nuts.

Providing Water

Just like us, birds have three basic needs: shelter, food, and water. Even though we live near a river that is a constant source of water, we like to keep a convenient source of fresh drinking water available, as well. Hey, if you ate dry seeds all day, you'd need an occasional drink too! Fresh water is as welcome to birds as your feeders are, particularly in the winter. Many areas don't have the luxury of a large body of

ABOVE: Birds are especially attracted to water movement. This little spray of water is the perfect enticement.

water nearby, so birds often have to rely on small pockets of water and little streams that may dry up in summer or freeze in winter.

All that is needed is a shallow tray that can be easily accessed by both small and large birds. In winter, that may be as simple as setting out a fresh tray of water on your deck every day. Or, if you have a larger birdbath, just make sure to keep the water fresh and open. Here in Virginia, if a layer of ice appears, it is a simple matter to break it or skim it off. But if you live in a colder climate, you may want to consider a simple water heater.

Problems and Misbehavior at the Feeder

Oh, yes. Whenever groups congregate, there are going to be issues, especially where food is involved. Posturing and territorial squabbles happen every day, and you may see some species dominate at the feeder. Larger birds will hog the seeds and eat too much, while some little birds will boldly dive-bomb other feeders. Usually everyone works it all out, but there are some situations that require special attention.

Knowing the pitfalls and potential issues can give you a head start on preventing many of them, but others can be an ongoing element of bird feeding. Here are some of the problems you may encounter.

Squirrels

Squirrels! The minute you decide to put out your first feeder, the drumbeat goes out. Hey, sunflower seeds—free for the taking! Come on, everybody! Not only do squirrels love the sunflower seeds and nuts that we put out for birds, but they are also clever and dexterous enough to wreak havoc on your feeders. If they are not climbing on top of them to empty the contents, they are knocking them down, chewing them up, and dragging them off. After all, they are cute little creatures, and they figure you were trying to attract them to your backyard. Little do they know that they are *not* the ones you were inviting to dinner.

There are three strategies you can employ to keep the squirrels from ruining your bird-feeding program: You can keep the squirrels from reaching the feeder; you can keep them from getting the seed they want; or you can give them their own food. Try the following strategies to keep the squirrels under control.

Thwart the Acrobats

Squirrels can climb, leap, and hang off just about anything. Just when you think you have figured out how to keep them off a feeder, they find a way to get at it. I watch them sometimes sitting near one of my feeders. You can just see their little brains turning. "How do I get at it?" "Can I make a jump for it?" "If I can just pull it down, it'll be all mine!" The reality is that squirrels don't have to actually reach the feeder. If they can just get into a position to shake it and spill its contents, they are satisfied. But there are a few things that will slow them down, if not stop them in their tracks.

- **Slipping and sliding:** Squirrels can't climb PVC, aluminum, or other slippery surfaces. They have no problem with tree trunks, metal poles, or long branches. Line your mounting pole or post with PVC piping or sheets of aluminum to keep them from scampering up and helping themselves. I have heard of people using a thick coating of Vaseline on poles, but I don't recommend it. The sticky substance can get on feathers and cause more trouble for your birds.
- **Make it jump-proof:** Squirrels' jumping range is good, but it's not infinite. Hang feeders at least 10 feet away from fence posts or tree trunks, or anything else that will give them a good leap at your feeder. Keep feeders at least 5 feet off the ground, too. And watch out for overhanging branches that they can leap down from!
- **Go fish:** Hang your feeders from fishing line or stretch the line horizontally and hang the feeder in the center of the stretched line. Add plastic soda bottles or anything else that might keep the squirrel from getting a good grip on the line.
- **Baffle them:** Install baffles above or below hanging feeders. These cones or cylinders stop squirrels from getting at the feeder by providing a block they just can't get around.

Squirrel-Proof Your Feeders

Bird-feeder designers have been matching wits with these critters for years. Here are a few ways to keep squirrels from getting your seed:

- **Weigh them down:** Try a feeder with weight settings that prevent squirrels from getting at the seed when they are on it. The squirrel's weight on the feeder actually closes the feeding holes. Larger birds like crows or grackles will also trigger this weight limit.

RIGHT: A sturdy baffle installed on a
feeder pole thwarts this squirrel's plans.

RIGHT: A sturdy baffle installed on a feeder pole thwarts this squirrel's plans.

- **Keep seed out of reach:** Use a feeder with an outer wire-mesh cage that keeps seeds out of reach. You would think that this style of feeder might keep the birds off, too, but many birds have no problem getting at the seeds or suet that the squirrels can't reach.
- **Give them a whirl:** For pure entertainment value, get a feeder that spins when a squirrel alights on it.
- **Chew-proof:** Avoid wooden feeders or feeders with plastic perches and seed ports. Squirrels can make short work of these if they get their paws on them. Their ability to damage wood and plastic is remarkable.

Take Away the Appeal

Squirrels want what everyone else is having. Sometimes the only answer is to feed foods that will make them turn up their noses. Either that, or just give in and give them their own!

- **Feed for who you want:** Attract only birds who like nyjer seed. Squirrels don't like nyjer and generally will not bother a thistle feeder, even if that's all that is available.
- **Bitter can be better:** Fill your feeder with safflower seeds instead of sunflower. While some birds may turn away, a lot of them will accept safflower. Unfortunately, even squirrels may become accustomed to the taste eventually.

- **Make it spicy:** Buy commercial "hot mix" suets that include hot pepper, or mix cayenne pepper into your own suet recipes. Birds do not have taste buds and won't be negatively impacted by the taste of the pepper. Squirrels, however, can taste the pepper, and they don't like it one bit. Some people mix dried hot pepper into their seed, too. There is a little controversy about this method. While the birds don't taste the pepper, it may burn on the way down. I personally don't use dried pepper directly in seed. Powdered pepper can irritate eyes and lungs—both yours and the animals'.

- **Just feed them already:** No one is going to expend energy they don't have to, so if you can't beat them, join them! We put a ground feeder at the edge of our forest and fill it with inexpensive cracked corn and a few nuts. Or you can buy corn stakes that hold ears of corn in place for them to gnaw on. The squirrels are happy for the convenient meal, and if they

ABOVE: This squirrel has a double problem—how to reach this weighted feeder, and how to get at the seed once it does.

come to view this feeder as "their" territory, they may leave your bird feeders alone.

In spite of your best efforts, squirrels may be an ongoing problem. Keep in mind, though, that the average life span of a backyard squirrel really is only a year or two. If you have one that is particularly pesky, you may not have to deal with him for more than a couple of seasons.

Cats

If you have a cat in the neighborhood, watch out! Cats view bird feeders as their own kitty buffet, and will stalk feeders for tasty songbirds. Domestic cats actually kill millions of songbirds every year. Here are some ideas to help protect your bird population:

- **Keep cats inside:** If you have a cat yourself, keep it indoors. It can enjoy the birds from your picture window.
- **Make cats noisy:** If you know your neighbor has an outdoor cat, ask them to put a bell on its collar to warn birds of its presence. When I saw new neighbors moving in with cats, I actually gave them cute collars and bells as a housewarming gift. The cats don't stray far from their own yard, but at least I know we will hear them if they do come around.

LEFT: Cats are always looking for their chance; luckily, this one is safely behind a pane of glass.

- **Choose safe locations:** Make sure that you place feeders in a safe place. There should be plenty of available perching and hiding places near your feeder so birds can flee if danger approaches.
- **Create obstacles:** Add stakes or ornamental hedge fencing underneath the feeder that will prevent cats from having a clear, safe landing area.
- **Remove the enemy:** If you consistently see a stray cat lurking around that clearly does not have a home, consider trapping it and delivering it to an animal shelter.

Raccoons

I am not sure there is anything more vexing than trying to match wits with a raccoon. You thought the squirrels were smart? Well, they don't hold a candle to the wily raccoon. This creature will not only empty feeders, terrorize nests, kill baby birds, and dig up your garden, but it will also stay up all night doing it, leaving you with a huge mess to clean up in the morning. Many of the same strategies you used for deterring squirrels are also used for deterring raccoons, although in my experience, with less success. They think faster, jump higher, and are stronger than any squirrel. Spring can be an especially bad time, especially if a female with babies has decided that you are going to provide easy meals.

Raccoons only come out at night, so if you suddenly notice your feeders are empty in the morning, don't blame the squirrels. You probably have a raccoon.

If you have hanging feeders, the best strategy may be to take your feeders in every evening. We did this one season for a relentless raccoon, only leaving one feeder up that the animal had never touched because of the way it was hung. But without the lure of the other feeders, the raccoon decided to tackle this one. The wily creature made short work of it, dropping it to the ground and eating its fill, leaving a broken feeder for us to clean up the next day. Raccoons have great memories too. Taking feeders down every day for a week won't really break the cycle. If the feeders are back up at night on day eight or day twelve or day sixteen, the raccoon will be there to take advantage of it.

An oversized trash can with a tight lid or an extra bungee cord can be kept near your feeders as your nighttime "lockbox." We put a concrete block on top for good measure!

After a season of doing battle with raccoons, we finally invested in a high-quality pole system with heavy-duty baffles. We position it well

RIGHT: Deer are generally not a problem at the feeder, but will compete with birds for fruit and berries.

away from any kind of jumping-off point. It held up pretty well under the onslaught of attention that the raccoon gave it early one season. Now, this is one feeder station that the raccoon leaves alone, apparently having decided that it's just not worth the trouble.

Other Mammals

When you offer food, you put out a calling card for everyone in the forest. This can include deer, as well as rats, skunks, and bears.

Deer compete with our birds for garden bounty and cherries off our trees. They will eat out of a bird feeder too, and they can stand on their back legs to reach your feeder. While we haven't had a lot of trouble with them at our feeders, they can definitely cause trouble in our shrubbery. On the other hand, our resident doe often bring her babies to visit, and we can't stay mad when we see adorable, awkward twins cavorting around the yard. We have a deer-free "safe zone" in our yard—a large, fenced-in dog yard that the deer don't enter, in spite of enticements such as the vegetable garden and their favorite flowers.

Bears are naturally curious, so if they are in the neighborhood, they may stop by to check things out. Human and bear proximity is not good for anyone, so if you have a bear, stop your feeding program and take in your feeders. Bears—especially females with cubs—are dangerous and unpredictable. The good news is that they will be gone by winter, so you should be able to safely feed again after December 1.

Skunks are not good climbers, but they are scavengers, so if you put something out that they consider worthwhile, they may become regular visitors. Be careful not to leave pet food outside, as it is a sure attractor for skunks and raccoons. Mice or rats can become a nuisance if they find a steady source of spilled seed or find a way to get into your stash. The best way to deal with both of these problems is good hygiene. Don't put out more feed than the birds can eat in a day, and rake up any spillage. It there are few seeds or scraps on the ground, there is not much enticement. Have good food storage routines.

It is very important to take note of any mammals that seem to be diseased. If you see odd behavior, an animal is not sturdy on its feet, or it's foaming at the mouth or unnaturally bold in your presence, immediately call animal control or wildlife management in your area.

Predatory Birds

Hawks and owls take the term *bird feeder* a little too literally for my taste. These hunting birds stalk squirrels, rabbits, and other small mammals, but unfortunately for our backyard guests, some specialize in hunting birds. This is just what they do, and from nature's perspective, it is no different than our little birds' insect-hunting instincts.

While I enjoy seeing a soaring, circling hawk and thrill to the sight of a barred owl in a tree, I really don't want them at my feeder. While some people may enjoy witnessing the thrill of a live-action chase, I have a soft spot for the birds that visit my backyard, and would rather try to help them out in a situation like this. The good news is that small birds already have some strategies for keeping hawks at bay. You may notice a sudden rise in urgent calls rising throughout the forest. Birds are good at spreading the word about danger. If you look around, chances are you will see a circling hawk. Feeding and traveling in flocks also helps provide some protection.

There are a couple of things you can do to help keep your birds safe. Provide good cover near the feeder so that the hunter cannot take direct aim at your birds out in the open. Likewise, keep your birdbaths in a sheltered location so your birds can enjoy the water in relative peace. A lot of hunting birds are opportunistic; they are just cruising by to see what's up. But if a hawk or an owl settles in for what looks like a long visit, you may want to take in your feeders for a few weeks. If it sees its opportunities dwindle in your backyard, it may move on.

It is usually relatively easy to identify the hawks at your feeder. The most common feeder-visiting hawks are the Cooper's hawk and the sharp-shinned hawk. The Cooper's hawk prefers medium-sized birds, while sharp-shinned hawks go for the smaller ones. Red-shouldered hawks are common in our area, and while we occasionally see them come through, we have never seen them stalk the feeder area. Common owls like the barred owl or the great horned owl may also note the plenty at your feeder, although

you will generally not see them in the same area; great horned owls will actually prey on barred owls.

Nuisance Birds

Some large birds can wreak havoc at the feeder, chasing away your pleasant little birds and acting like Great-Uncle Harry at Thanksgiving dinner. Platform feeders are attractive to these birds because they are easy for them to access. Likewise, feeders with large, sturdy perches make food pretty accessible. If the ground is covered with seed, they also consider it an open invitation to dine. Here are some of the most common problem birds:

European Starlings

It is particularly dismaying to see European starlings descend upon your yard. Traveling in large flocks, they can clear out a feeder in no time, chasing away your regular guests. If they set up to roost, it may be hard to get rid of them. Some people put up owl decoys and use predator calls, but it doesn't take long for starlings to realize that there is no real danger there. If you don't want starlings in your yard, immediately go on an aggressive campaign of chasing, shooing, and annoying them. Be persistent. It's not easy to convince the starlings to leave. Use loud noises to flush the birds from the trees, and do it every day. Unless your environment is particularly compelling, they may just decide that the crazy lady's feeders are just not worth the trouble!

Other Birds

For one reason or another, you may find some species less desirable at your feeder. It may be that they come in large flocks that overrun your yard; it may be their aggressive attitude; or it may be that they have unsavory habits that you don't want to encourage. Some of these birds are larger—crows and grackles come to mind. These birds can be discouraged at the feeder by using styles that don't accommodate their weight or perching needs. Feeders designed to shut down with the weight of a squirrel will do the same thing for a heavy crow. Bottom-feed suet feeders prevent large birds from getting at suet. There is often something to eat on the ground below a feeder, so if you notice an increase in large flocks of undesirables, it's probably time to rake up spillage.

Some birds, like house sparrows and cowbirds, are often unwelcome because of their nesting and marauding habits. House sparrows will displace other birds from their nests, and cowbirds have the outrageous behavior of placing their eggs in other birds' nests for others to hatch and rear. Although I enjoy my clever blue jays, some people would rather not see them either. They consider them noisy and pushy, and accuse them of marauding other nests.

Wasps, Bees, and Ants

Hummingbird feeders can be the target of wasps, bees, and ants, all of which love the supply of sugary nectar. Cleaning and replacing your nectar weekly will help keep this problem to a minimum, but if you have continuous issues, consider a feeder with an ant moat. This moat contains water that ants can't cross. There are also feeders with bee guards that prevent them from reaching the nectar. I have the best luck with saucer-shaped feeders because they have top feeding ports. Hummingbirds can reach the nectar inside, but insects can't.

LEFT: The clever blue jay is alert to every opportunity and will watch for his chance to take advantage of a situation.

You can see that as your feeder program grows, you may begin feeling like you need to wear a wildlife manager's hat! It's all about balance in your backyard. After all, you did invite the birds to come, remember? They can't help it if you want to play favorites. A solution for one problem may result in other problems popping up. That is the beauty of nature, and the danger of trying to alter it with your own actions. Remember my suggestion to give inexpensive corn to the squirrels? Well, crows, grackles, and cowbirds also love corn, and may take you up on your offer, too! A new problem! Hate those hawks and raccoons? Get rid of them, and there's no one to pick off the mice and other pests that may be bothering you. Pick your battles, make accommodations as needed, and enjoy the pageant of wildlife parading by your window.

> **"** The tables are set and I await my feathered guests. **"**
>
> **—DAVID M. BIRD**

⑤

Recipes and Blends for Your Feathered Friends

FOR THE most part, birds are perfectly happy with the basics. But there are some special treats that may attract hard-to-entice species or provide an extra boost for them during nesting and the cold season. We are especially mindful of our birds during the holiday season. Just as we are sharing our abundance with loved ones, we like to remember the little creatures outside our door who are just beginning a long, difficult season.

It is helpful to know what kinds of birds like what types of foods. My mother used to throw out scraps of bread, knowing that someone would eat them, and in fact, a lot of birds enjoy bread products. But others have no interest whatsoever. Keep the following list in mind when offering food from your own cupboard:

- **Bread and grains:** A lot of birds like bread, cornmeal, millet, and oats. Blue jays and crows are always ready to grab for a meal of bread. And other birds are perfectly content to have bread crumbs, oats, or cornmeal mixed into their food. Wrens, mockingbirds, thrashers, sparrows, warblers, titmice, cardinals, grosbeaks, buntings, and chickadees will stop by for this treat.

- **Fat:** Fats are a crucial source of fuel during nesting and in winter weather. Suet is the first ingredient that comes to mind when we think of birds. Suet is made from raw beef fat. It is a favorite of nuthatches, woodpeckers, wrens, titmice, chickadees, and cardinals. Even blue-birds will occasionally stop by for a taste. I keep both lard and suet around for making mixes, and stir in leftover bacon grease for a special treat. Other oils, like olive and corn oil, can also be added to mixes.

- **Peanuts and peanut butter:** Many birds share our enjoyment of peanut butter, but a lot of the brands that humans like are full of sugar and additives, so I keep a jar of natural-style peanut butter just for them. I use peanut butter in a number of my mixes. Just about every bird that likes nuts, seeds, and fat will come for a taste of peanut butter. Chopped peanuts have become more accessible, and even cardinals and finches like the taste.

- **Dried fruits (i.e., dried apples, raisins, and currants):** Dried fruits are the perfect replacement for birds who like to forage for berries during the regular season, including waxwings, orioles, and thrushes.

- **Sunflower seeds:** The sunflower seeds in your cupboard are probably the old-fashioned striped sunflower seed. But black-oil sunflower seeds have become the premium food of the bird world. They are high in fat, and wild birds from jays to finches enjoy the taste. The shell is slightly thinner than the old-fashioned striped sunflower seeds, and they have larger kernels. So while your birds won't mind a taste of your sunflower seeds, they would probably prefer the black-oil seeds that you keep just for them.

- **Apples:** Apples are an easy-to-keep fall and winter fruit for us, so it's pretty easy to offer fresh apples in mixes, or cut up and left at the feeder. Because birds eat early and late, it isn't necessary to leave fresh fruit out all day. Put it out at breakfast and again at dinnertime, and refrigerate it in between. A lot of birds enjoy fresh fruit in the winter, including waxwings, mockingbirds, thrashers, wrens, cardinals, grosbeaks, and buntings.

- **Oranges:** To me, a fresh-cut orange left on the feeder seems like one of the most hospitable sights of winter, and a number of birds will come for a taste, including orioles, finches, tanagers, and even woodpeckers.

Place half an orange onto a large nail that is hammered into a stump or fence post, or position them on platform feeders or even in suet cages.

Recipes

Everyone has a favorite recipe for bird-food and suet mixes. Most share the same basic ingredients. Some recipes work especially well in certain locations and with certain birds, so share your successes with your neighbors.

Suet from Scratch

I generally don't make my own suet, but I have friends who say their birds want nothing but their special recipe! Suet is just fatty beef tissue. You can buy beef fat at the grocery store, or just save the trimmings from your own beef and freeze it until you have enough for rendering.

Grind the beef fat with a meat grinder (a butcher may be willing to do this for you). Heat the fat over low heat until melted. Strain the melted suet through fine cheesecloth. Cool until firm, and then repeat the melting and straining process one more time. Rendering the fat twice will allow the suet to form into firm cakes. Chill until firm. Your suet is now ready to use in other recipes.

For suet cakes, pour into molds and chill. Square plastic sandwich containers are the right size for many suet cages. Store in freezer.

Summertime Suet

Birds appreciate suet all year-round, but if the temperature is above 60° Fahrenheit, beef fat will turn rancid and melt, staining decks and pavement, and making a sticky mess of things. I don't put out much suet in the summer, but during spring nesting season, it is very much appreciated. In my area, spring days can be very warm, and this recipe works very well.

I cup crunchy peanut butter	2 cups cornmeal
I cup lard	I cup white flour
2 cups quick-cook oats	¼ cup sugar

Melt the lard and peanut butter over low heat in a large pot. Stir in the remaining ingredients. Pour the suet mixture into square plastic sandwich containers, about 1 ½ inches thick. Store in freezer. These fit into your standard suet feeders. Makes three sandwich-sized loaves.

Fall Migration Mix

Many birds feed primarily on fruit, so as supplies dry up and bushes get picked clean, they start to think about moving on to new locations. This mix can help them along their way. It makes the most of seasonal fruits and is a tasty snack for migrants.

Mix together:

I cup dried apples, chopped into small pieces	I cup walnuts
I cup raisins	½ cup dried pumpkin seeds

Ground foragers like this mix, so feed this recipe on a platform feeder. Store the mixture in a paper bag to keep it from getting moldy.

Classic Mixed Bird Food

A basic blend with a little something for everyone. A food processor makes this recipe a lot easier. Make sure your storage container is clean and dry, with a tight-fitting lid.

4 cups of shelled black-oil sunflower seeds, pulsed quickly in your food processor to a coarse chop	I cup of cornmeal
2 cups of whole oats, lightly ground in your food processor	I cup of dried fruit, chopped into tiny pieces (almost any dried fruit will work, including raisins, apples, dates, or apricots)

Pour all the ingredients into the bucket and mix. I reach in with my hands and toss the layers of ingredients together; or, you can shake the bucket—although this is easier to do if you add just a couple of ingredients at a time, rather than all at once.

Nuthatch Blend Suet Log

Flickers, nuthatches, and chickadees flock to our suet log, made from a 6-inch-diameter log, cut about 12 inches long. Make sure to leave the bark on for easy footing, or add wooden-dowel perches. Add an eye-hook for easy hanging from a tree.

I pound of suet	I birch log with several shallow 1½-inch holes
I cup each of cornmeal, oats, chunky peanut butter, mixed birdseed, and hulled black-oil sunflower seeds	

Melt suet in a large pot over low heat. Stir in dry ingredients and let cool. Pack suet mixture into holes in log.

Passerine Power Bars

On a cold, snowy day in January, there is nothing better than a quick fat-and-carb bite to get the day off to a good start. Keep these protein-loaded bars around all winter for your winter residents.

I cup fat—a combination of lard and bacon fat works well	¼ cup molasses
I cup chunky peanut butter	I cup whole-wheat flour
2 cups cornmeal	¼ cup raisins
2 cups oatmeal	¼ cup hulled sunflower seeds

Melt lard and peanut butter over low heat in a large pot. Stir in the remaining ingredients. Pour the mixture into an 8x8-inch square pan. Refrigerate until firm, and then cut into small squares. These freeze well. I put them on a cookie sheet so they can freeze in individual cubes, and then transfer them into Ziploc bags for storage.

ABOVE: Northern flicker

Ground Feeder Winter Suet Blend

Ground foragers like cardinals, jays, and grosbeaks have a hard time with those small swinging suet feeders, so I keep this crumbly blend on hand for them all winter long.

I cup suet	4 cups cornmeal
I cup chunky peanut butter	I cup whole-wheat flour

Melt suet and peanut butter together over low heat. Combine the rest of the ingredients together in a large bowl. Cool the fat mixture slightly and then stir into the bowl. Mix thoroughly into a crumbly mix and serve on platform feeder. Freeze for use all winter long.

Christmas Brunch Bark

I make an elaborate breakfast during the holidays as a way of showing my appreciation to my birdie friends. I set aside a handful of ingredients from my own holiday baking to add to their holiday treat. I put some in the suet feeders for the small birds, as well as on the platform feeder and on the ground for the cardinals and doves. Of course, rabbits and squirrels come too, but hey—it's Christmas!

3 cups fat (can be a combination of suet and bacon fat)

1 cup bread crumbs

½ cup black-oil hulled sunflower seeds

¼ cup chopped pecans

½ cup chopped dried apples

¼ cup raisins

¼ cup chopped dates

¼ cup freeze-dried mealworms (optional)

Melt suet in a saucepan over low heat. Combine the rest of the ingredients in a large bowl. Allow suet to cool slightly, then stir into the bowl of dry ingredients. Mix thoroughly. Pour onto cookie sheet and refrigerate until hardened. Break into pieces.

Holiday Nuggets

These tasty nuggets make a nice gift for birding friends. After you chill them, you can pack them together. I like to use Chinese take-out–style boxes—just make sure your label says they are for the birds!

2 pounds suet

2 cups chunky natural peanut butter

¼ cup bacon grease

¼ cup molasses

4 cups oatmeal

1½ cups cornmeal

Good-quality mixed bird feed

Melt suet over low heat. Pour into a large bowl and stir in peanut butter, bacon grease, molasses, oatmeal, and cornmeal. Add enough bird feed until mixture is firm and can be rolled into balls. Shape into 1-inch nuggets and roll each into the bird feed. Chill on a cookie sheet. Keep refrigerated.

Hummingbird Nectar

Hummingbirds and orioles both enjoy nectar. Use plain sugar, never honey or other sweeteners. Red dye should not be added. Change nectar weekly, and more often in hot conditions. Wash and rinse the feeder thoroughly each time you change it. Choose a shady or sheltered place for this feeder. Direct sunlight will cause the nectar to spoil more quickly.

¼ cup sugar	I cup water

Bring sugar and water to a boil to kill any bacteria or mold present. Cool, and fill feeder with the nectar. Store extra in the refrigerator.

ABOVE: It may take hummingbirds a while to notice your offering, but in the end, you may be rewarded with loyal customers.

Blue Jay Pie

We make this pie especially for our winter blue jays. This is a great treat to make with children. Gathering and adding a handful of chopped acorns help children to understand where birds get their food in your yard. We place this pie right on our deck railing so we can enjoy watching the jays.

1½ pounds suet

½ cup crushed peanuts

½ cup black-oil sunflower seeds

½ cup cracked corn kernels

½ cup chopped acorns (optional)

Melt suet in a saucepan over low heat. Stir in nuts, seeds, and corn, adjusting amounts to make a thick blend. Spoon the whole thing into a foil pie pan and sprinkle the top with birdseed. Cool in the refrigerator until solid. Put out on platform or ground feeder.

ABOVE: Blue jays come every day to see what's on the menu at our house.

Dried Bird Apples

A farm near my house has an old, abandoned apple tree whose fruit is a boon to the local birds all season. Unfortunately, the apples are not great for human consumption—too many wormholes! But what could be better for wintertime bird treats? I make a special batch for my birds and feed it to them all winter.

Pick leftover late-season apples. Wash and slice into ½-inch slices. Give them a quick wash in lemon water to prevent browning. Place apple slices onto drying rack. Dry fruit in oven at low heat, about 150 degrees Fahrenheit, overnight. When apple slices are dry and pliable, they can be stored in airtight bags or jars. After proper drying, apple slices will keep in a cool, dry place from six months to a year.

BELOW: A cedar waxwing tastes to the last of the winter apples.

> " I value my garden more for being full of blackbirds than of cherries, and very frankly give them fruit for their songs. "
>
> **—JOSEPH ADDISON**

⑥
Your Backyard Sanctuary

ATTRACTING BIRDS to your yard can be as simple as putting up a feeder. Bird feeders can be a great supplement to their natural forage, but they can't replace it. If you want to really learn about the lives and habits of the birds in your area, examine your yard from their point of view. The reality is that as humans, we are an inescapable part of the natural world, and what we do with our small corner of it can make a huge difference to the bird population.

For the most part, we follow the concept of "wildscaping" in our yard. Wildscaping provides an environment that is friendly to birds and wildlife by expanding native plant life into areas of human habitation. We plant and encourage native vegetation, install nesting boxes, and provide water sources. As a result, we are rewarded with singing birds, nesting parents, adorable fledglings, and the knowledge that our yard is a healthy and useful part of the planet!

LEFT: House sparrow

Assessing Your Yard

Take a walk in your yard and think about it from the perspective of a potential bird resident. Just as you would evaluate your home for warmth, convenient location, good schools, and safety, birds also search for an area that can fill their needs. These needs are quite basic. Like every living creature, they require food, water, and shelter.

With such a large variety of birds in the world, their food requirements vary considerably. Food needs range from flower and weed seeds for finches to berries and fruit for orioles and blackbirds, from worms for robins and other thrushes to insects for the flycatchers and martins. The list goes on and on.

Besides your feeder, is there anything in your yard that naturally attracts birds to stay and eat? Do you have a big evergreen tree? Besides being a great nesting location for certain birds, it can also provide a great source of food for varieties that like the seeds that come from the pinecone. Pine siskins, titmice, chickadees, and other birds love these nutritious seeds. The blue jays will thank you for that big oak tree—they love acorns! Decorative bushes and shrubs can be great hiding places, but are yours also a source for berries and nectar? Your lawn can be a great location for spring and summer worm foraging, but is it bird-healthy or full of pesticides? Your vegetable and flower beds, your fields and your woodlot can all prove beneficial to the bird population and provide an enticing place for migrating birds to settle for the season.

Make note of the good local plants and trees that you have. Get some graphing paper and draw out your yard in its current state. When we bought our house, it was blessed with lots of azaleas and rhododendrons, as well as a wealth of native holly bushes and dogwoods. It had ivy groundcover that was great for the local wrens and sparrows, and a

LEFT: This American robin appreciates a berry thicket almost as much as the earthworms in your lawn.

nice woodlot with a handful of dead trees that were perfect for woodpeckers and nuthatches. We were very lucky—there was a lot to work with.

Many yards contain two extremes—tall trees, and a wide expanse of lawn. In spite of the fact that we had plenty of natural growth and some great landscaping near the house, we soon saw that our woodlot was very mature and needed more undergrowth. The lawn was gorgeous but had long been chemically maintained. And the place definitely needed more fruit and berries!

Enhancing Your Environment

After you have evaluated the major bird features in your yard, take a look at the potential locations for improving the environment. Learn who your local birds are and what they like. If you are planning a landscaping project, select native trees and shrubs rather than exotics. They are going to be better suited to the local bird population, and are generally hardier than exotic choices. Coming from the North, we did not immediately recognize the Virginia creeper in our woods. We thought we had a major poison ivy crop! Once we learned that it was actually a "good neighbor," and that over thirty-five species of local birds eat its berries, we let it spread and take over part of our fencing. Your local extension service can be a great source for providing recommended native nursery choices. Choose bushes with flowering berries and trees with nuts or berries. Think about humming-birds when choosing your flowering plants, and consider a mixed hedge for lawn borders.

We are fortunate that our land is adjacent to a large tract of undeveloped forest, as this greatly expands the natural attractiveness and breeding range of our area. It makes our lot an ideal location for wildscaping. But if you are in a newly established neighborhood, or just getting started on a new lot, this method of landscaping is even more important. Remember: Your new subdivision used to be home to all kinds of native wildlife that may have been driven away by bulldozers. Ask your builder to retain the native trees, grasses, and soil; they will serve as a great foundation for your new bird-friendly yard.

When we bought our four-acre property ten years ago, our first step was to reduce the size of the lawn by adding more hedgerows and flowering bushes. We expanded the natural groundcover and sowed a hard-to-mow hillside with wildflower seeds. Then we began to work vertically, adding short trees under the towering tulip trees and expanding the understory by planting native hollies, rhodies, and azaleas further into the woods. We established flower beds with hummingbirds in mind, and added fruit trees that we were pretty sure we would never get to harvest ourselves.

We had some hard choices to make, too. A dead oak threatened our roof so it had to go. We had to remove a few shrubs that were threatening our foundation. We tried to make up for any losses in our yard by planting other trees and shrubs in more-suitable places. Our kids require a certain amount of lawn, but we have found ways to maintain it without pesticides. It's a little weedier than the average showcase lawn, but it's also healthier.

ABOVE: An assortment of native trees and shrubs provides a beautiful landscape and lots of cover for birds.

Our yard is an ongoing project and one that we find continuously rewarding. We can't wait for our first visit to the nursery each spring.

Bird-Friendly Habitats

Spring is a busy time for birds. They are traveling from their winter grounds, going through their prenuptial molt, and getting ready for mating season. They are starting to search for nesting areas and preparing for their courting rituals. As the weather warms up, they may start to abandon your feeder in favor of fresh insects and new nesting territories.

Most birds are looking for a good source of insects in the spring, so watch for your birds to be on the hunt. Most birds that have been loyal to your feeder all winter are anxious to get out and find some fresh protein. Caterpillars are a favorite of your cardinals. Bluebirds and towhees love beetles, grubs, and spiders. Warmer weather brings our first outbreaks of mosquitoes and other flying insects, which will put warblers, phoebes, flycatchers, and cedar waxwings on the prowl. As soon as the ground warms up, you will quickly note the arrival of robins in search of earthworms. The burst of new green leaves brings out other insects on branches and in bark. Look for busy chickadees, titmice, nuthatches, and woodpeckers to be busy there. Open water offers up larvae and mayfly hatches for a wide range of birds. Don't like bugs yourself? Never fear! A healthy yard full of insects always attracts birds that are glad to keep the place safe for you.

Planting

Phase One: Foundation Planting

While your birds are out hunting, now is the time to do a little planting. My plan for spring planting includes two phases, and phase one is foundation planting.

Where in the yard am I going to invest in a new tree, some new shrubs, or some perennial berry bushes? Budgetary considerations require me to make just a few such purchases a year, so I take inventory of my yard to decide what I want to add. Usually my husband and I have dreamed up projects over the winter and know exactly what we want to do, so it's just a matter of prioritizing. But if you are new to wildscaping, consider the following:

Hiding places: Your birds need natural shelter, so think about creating more hiding places for them by adding berry bushes and flowering ornamentals. Look at travel routes around your yard. Is it easy for birds to travel under cover, or are there wide-open spans of lawn? A mixed hedge of shrubs makes for easy travel, as does the addition of some shorter trees near tall ones. Shrubs are also ideal safe nesting sites for wrens and other small birds that like lower locations. Choose barberry, hawthorn, lilac, privet, or any number of a wide array of berry-producing bushes.

Shrubs: If you were planning to buy shrubs anyway, look for natives with food-producing capabilities. It is not hard to please both yourself and the birds. The amazing burning bush produces a fantastic pop of color that I love, and the birds really go for the bright berries that come along with it. Beautiful evergreen hollies are native to our area, and the birds love the red berries they produce late in the season. But don't think a single specimen bush is going to make the ideal bird home. A lone shrub in the middle of the yard says "Look here!" to predators. Variety is the key. Try groupings of three to five shrubs together to make a pleasing corner of the yard that is also bird-friendly.

Layers: Think in layers; find ways to fill the vertical space between that tall oak and your lawn. Some small trees, like dogwood, adapt naturally to that shady middle space. Shrubs add the next layer down and perfectly suit small birds that like to nest only 3 to 5 feet off the ground.

Go large: Tall trees such as ash and tulip provide plenty of seed. Oaks are great for acorns. Large evergreen trees such as juniper, spruce, hemlock, and fir offer great shelter in bad weather and provide food for cone-seed-loving birds.

Consider the middle ground: Shorter, decorative trees like dogwoods and smaller evergreens grow in the shade of taller trees and produce plentiful fruit in the fall. Fruit trees such as cherry, Juneberry, mulberry, and crab apple are always popular with the birds. Consider

LEFT: Even the more-domesticated parts of our yard provide a safe travel route for our birds. Although not strictly a native, the dwarf Japanese maple in the foreground was in place when we arrived, and at thirty years old, has earned its place in our yard. Evening grosbeaks enjoy the seeds.

LEFT: This corner of the yard was a desert of empty space between the lawn and the treetops before we added native flowering dogwood and azalea.

planting a hedgerow of mixed tall shrubs with fruits and seeds that mature at different times. Sumac and dogwood are a great source of fall food; rosebushes offer berries long into the late seasons; and holly and hawthorn provide berries in the winter months.

Low and lovely: Berry shrubs (like blackberry, raspberry, blueberry, grape, and strawberry) and evergreen shrubbery (like rhododendron, yew, and spirea) are great for nesting birds. Berry-bearing autumn shrubs like winterberry, spicebush, sumac, and Virginia creeper are welcome after the summer fruit is gone. Hummingbirds love azalea, butterfly bush, and weigela, among other flowering shrubs. Shrubs that offer berries into the winter include hollies, barberry, sumac, and roses.

Ground play: Don't forget the bottom layer of your yard. For a partially sunny area where no one walks, consider adding vines. A tangle of honeysuckle will thrill the hummingbirds, and a stand of Virginia creeper will

BELOW: A stand of native grasses and vines softens the edges of our yard and provides seeds and berries to the local birds.

attract all manner of birds. Bittersweet and wild grapes are also great choices. And that crop of poison ivy? Birds actually love the white berries they produce, so if you find an out-of-the-way patch, you may want to note that area for bird-watching!

Phase Two: Flowers and Vegetables

Phase two of spring planting is flowers and vegetables. Keep your birds in mind when you choose your flowering plants.

Flowering plants: A garden filled in springtime with red and orange nectar-producing flowers will attract hummingbirds later in the summer. It will probably attract butterflies, too, which may in turn attract flycatchers and other birds seeking flying insects.

Think color: Choose red salvia, columbine, honeysuckle, coneflowers, and fuchsias for hummingbirds and other nectar lovers.

Annual investment: The annuals that you plant are not only good for providing spot color throughout your garden and planters, but their extended blooming time also provides seeds and nectar long into the season.

Go wild: Sparrows and finches make the most of wildflowers and weed seeds, so if you have room, consider a corner of your yard for assorted wildflowers. Some wildflowers that thrive include goldenrod, yarrow, coneflower, joe-pye weed, and black-eyed Susans.

Attract bugs: Plants like sedum and goldenrod attract insects, which in turn attracts insect-loving birds.

BELOW: Native flowers and grasses provide a wealth of weed seed to summer nesting residents.

Building fund: Willow and milkweed produce food and nesting materials for weed-eating birds.

Cultivate late bloomers: Late-blooming flowers like salvia, hollyhock, and lobelia are greatly appreciated by hummingbirds as they start their fall migration.

Fruit and vegetable gardens: As to our own fruit and vegetable gardens, we are willing to share—to a degree. If you are trying to grow your own food, birds are a good news / bad news situation. Sure, they help a lot when it comes to natural pest control. They are happy to cruise your beds for insects that might be eating your leafy greens or chomping on your tomatoes. But they also have a bad habit of picking up seeds you just planted or helping themselves to juicy strawberries. (Okay, I wouldn't mind sharing, but couldn't they just take the whole berry rather than hopping along, putting holes in each one like the visiting blackbirds do?) To mitigate the problem, we have a few strategies:

1) **Skip the seeds:** We use bedding plants instead of seeds for as many of our vegetables as we can. That way we don't lose whole rows of plantings before they even get a chance to germinate. Start your seeds indoors to get a jump on the birds.

2) **Hire security guards:** Our dogs cruise the area where the vegetable garden is. They are pretty good at keeping marauding flocks of crows or blackbirds from making themselves too comfortable. And while they squabble a bit with the birds, they have never hurt one. If your dog is an aggressive bird hunter, you may not want to try this one.

3) **Provide alternative food:** We also grow a few sunflowers specifically for the birds. We keep a feeder nearby. I think we may be kidding ourselves here. After all, if you were given a choice of fresh fruit or the same seeds you have been eating all winter, which would you pick? But it makes us feel that we are trying to be fair.

ABOVE: Security guards keep our vegetable garden safe from flocks of blackbirds.

LEFT: Late-season strawberry beds are fair game for the birds.

4) Create a barrier to entry: We put bird netting over the parts of the garden that we really don't want to share. At least, we try to. Every May, when the cherries start to form, I say to my husband, "Time to put up the netting." I say this every day for a few weeks. Then one day, we each reach for the first bright, almost-ripe cherries. Delicious. Okay, that net has got to go up tomorrow! As soon as we make that resolution, however, the birds decide the berries are ready. Inevitably, they clean the tree of every piece of fruit by the very next day. No cherries for us . . . again. But we know what a thrill those trees are to our birds.

5) **Raise a ruckus:** Scarecrows, shiny pie plates, or wind chimes can be used to ward off birds. Think shiny, reflective, and noisy. Flapping banners, streamers, or flags only work in the short term, as birds tend to get used to them. We gave up these methods long ago. Our birds just seem to find them amusing!

6) **Harvest promptly:** Birds have a good eye for fresh food, so if you want your fair share, try to take a walk through your garden and harvest every day. Overripe vegetables attract insects that bring more birds into the garden, and can lead to them investigating plants that wouldn't be interesting otherwise.

7) **Share the wealth:** Chop overripe melons or squash in half before throwing them on the mulch pile. Birds that like the seeds will be happy to find them. At the end of the season, leave some apples or berries behind for the birds, and let old crops stand for fall and winter foraging.

Water Sources

Birds need water for drinking, of course, but many also need a place where they can "freshen up" now and then. It is important for birds to keep their feathers in good condition, both for flying and for maximum heat efficiency when they need to stay warm.

When birds come to bathe, they fluff up their feathers, splash their wings to create a shower, and then shake off the excess water. This process is usually pretty quick—no long soaks in the tub! In fact, birds need less than a couple of inches of water to complete this process, which is why traditional birdbaths are wide and shallow. If you can watch your bird after he leaves the bath, you will probably be able to see him complete his preening process. He will go over each feather, seemingly combing it out or nibbling at each one until he is satisfied that he is back in good shape. He may hold his wings out to dry them.

Birdbaths

Safety is the primary consideration in choosing a location for your birdbath. Choose a lightly shaded area if possible, both to provide cover and to keep the water at a reasonable temperature. Birds will want to leave the bath rather quickly, but being wet, they probably don't want to have to fly too far, so make sure there is a quickly accessible perch or low-hanging branch nearby. Keep the bath about 3 feet off the ground.

Cleanliness is the next consideration. Your birdbath needs to be cleaned regularly to keep the water fresh, to remove droppings and parasites, and to prevent the spread of disease. Over time, I discovered that I am *not* one of those people who remember to go out and maintain areas that are far from the house. So I place my shallow trays and birdbaths in flower and vegetable gardens that get watered regularly. Because we need to water often, it is easy to hit those with the hose to freshen them up. When I deadhead flowers or pick ripe vegetables, I will occasionally give them a quick cleaning with a scrub brush set aside for the task.

My birds' favorite is an elevated birdbath tucked under a small dogwood in my front garden. It is a safe place for birds to take a quick sip and have a nice splash away from the watchful eyes of cats and

LEFT: A shady location near a sheltering tree is the ideal location for a birdbath.

other predators. And the garden gets watered regularly, so it is easy to keep it full and fresh. If you refresh the water often, you can get away with doing thorough scrubbings with a mild bleach solution once every season. Rinse thoroughly after using bleach.

Choose a birdbath that has a variety of depths. It should be about 3 inches deep in the center and have sloping sides that are shallower. Small birds need ½ inch or less for bathing, so you may want to add stones to create a shallower area. The edge should be broad and flat, and made of material that is easy to grip and not slippery. I don't care for terra-cotta baths for that reason. Although the glazed surface is pretty, there is not much to grip if the bird needs to make a quick getaway. Likewise, metal and copper can be too slick and can get hot if kept out in the sun.

Check your birdbath often to make sure it has water in it. Birds naturally splash a lot of water out of the birdbath when they bathe. My birdbaths also lose water to evaporation very quickly in the summer, so I often refill two or three times a week.

Fountains and Moving Water

Like us, birds love the sound of moving water. A trickle of running water or the gurgle of a small stream is a natural attractor. If you are a fan of fountains and have the interest and inclination to keep them clean and fresh, and free of leaves, droppings, and algae, then they are a wonderful source for birds. They do require maintenance, though, so keep a scrub brush or scouring pad handy and refresh the water often. Remember, additives that are traditionally used for keeping your fountain water clean are not healthy for birds that may be inclined to stop by for a drink. Some people are comfortable adding a small amount of bleach to fountains, but since this is a source of drinking water for the birds, I would rather not use it. It requires cleaning a little more often, but I think it is worth it.

If you are so inclined, there are also water wigglers, drippers, and misters on the market for making your water feature or birdbath more interesting.

Your yard sprinkler is an attractive place for birds, too, so don't be surprised to find robins and other birds hopping through the water, enjoying the sudden "rain shower," and picking off earthworms coming to the surface.

RIGHT: Although we would love to build a spectacular water feature in our backyard, our birds are happy with the small natural stream that runs through the property.

Natural Ponds and Pools

Waterscaping has become a popular form of landscaping in recent years. We love to see the beautiful and elaborate waterfalls, ponds, and fountains at our local garden expos, and sometimes wish that we could add a spectacular feature like this to our own yard. But for the bird-friendly backyard, these features should be maintained with a minimum of chemicals and pesticides.

If you want to build a natural pond that your birds can enjoy, take a little time to read up on pond ecology so that you can understand the delicate balance that goes into a healthy pond. A well-balanced water ecosystem requires good bacteria, phytoplankton, protozoa, and algae to serve as a natural bio-filter for the water. The zooplankton eat the algae and protozoa. The insects, larvae, and snails that live there will consume the zooplankton. Frogs, fish, and salamanders complete the roster of full-time pond residents. And finally, here come the birds! Birds may come to fish, to eat larvae and small insects, or for a quick bath.

A healthy pond requires a balance of all these elements. Snails, tadpoles, and other small scavengers can help keep the water clean, and the addition of certain plants will also help to do the job. A properly designed pond should be able to maintain itself naturally, but if you have to help to keep the ecology in balance, there are bacterial additives as well as aerators and other natural structures that can help to keep your pond clean.

Dust Baths

Some birds don't bathe in water at all, actually preferring a quick roll in the sand to maintain their feathers and to remove excess moisture or oil. You may have noticed a robin or sparrow scratching around a bare spot on your lawn, tossing up loose dust. They may even get down and roll in it. This process seems to be aimed at covering all their feathers with a light coating that they can then shake off. It is thought that it also helps to control mites and other small parasites. Maybe it just feels good, too, kind of like when we exfoliate our own skin.

You can assist birds with this bathing process by adding a small dust bath on your property. Rather than let the birds choose a spot, you might want to just take an area in your yard that already has loose soil and scrape it up a bit to make it more attractive. Just make sure it is in the sun, as birds don't really want to use damp soil.

A more formal dust-bathing area is easy to build. Choose a site that is sunny and safe, about 4 feet in diameter. Dig an area out to about 6 inches deep. You can get as elaborate as you want at this point. If you don't want weeds to grow in the dust bath, you can line the hole with a landscaping weed-proof liner. To keep the sand in and make it an attractive element of your yard, you can trim the perimeter with bricks, stones, or other edging. Then fill the bath with fine sand and mix it with the loosened soil.

Birds leave behind droppings, feathers, and mites, and unfortunately, cats and other mammals may want to make use of this area as their litter box. To maintain the dust bath, rake it occasionally to keep it clean. And don't let children use this area as a sandbox. This area is strictly for the birds!

Nesting

Some birds have very little interest in your feeders. Their primary concern is to find a clean, dry place to raise their young. While a variety of shrubs, hedges, and shade trees offer a lot of great potential building sites for many, some birds need a home that can be harder for them to find.

Humans and birds have had a long relationship when it comes to nesting. We have been providing nesting sites to purple martins since the Native Americans started hanging gourds for them. In fact, a large percentage of martins now rely almost exclusively on human-built martin houses for their shelter. Likewise, a concerted effort to establish nest boxes has resulted in an increase in the eastern bluebird population. As we clear more land for development, birds that require natural cavities suffer from the loss of those trees. Dozens of species of North American birds can be attracted by offering nesting boxes.

Nesting Boxes

You have undoubtedly seen any number of adorable birdhouses on the market or at local craft fairs. But not all birdhouses make great bird homes, so evaluate these houses based on criteria that birds would consider. You are attempting to accommodate cavity-dwelling birds, and there are over eighty North American birds that rely on such cavities. Here are some basic considerations:

Material

First, what is it made of? Birdhouses should be made of wood. This is the material that birds are accustomed to, and it provides the best insulation in both hot and cold weather. The outside can be decorated or painted, but the inside needs to be raw wood, preferably with a somewhat rough surface, so birds can make an easy exit without having to scramble on a slick surface. Avoid resin-impregnated "wood" that doesn't breathe, as well as terra-cotta or plastic houses.

Entry

The entry hole for most nesting boxes should be 1 ½ inches in diameter. A smaller hole may prevent common birds from using the box. A larger hole may invite marauding birds. Some houses designed specifically for wrens have a 1-inch opening, but unless you are very specific in what you are trying to attract, stick with the 1 ½-inch opening. Choose a box with a roof overhang that provides protection from the elements.

Size

Your box should simulate a hole in a tree, so the hole should be about 5 to 6 inches above the floor of the box. Likewise, the size of the box should be cozy: 5x5 inches is an ideal size for a cavity nester to build their nest.

Other Features

Your nest box should provide drainage and ventilation through holes or a small slit. It is also necessary for you to be able to open the box for cleaning and monitoring of the nest.

Location of Your Nesting Box

Nest boxes should be established before mating season begins. We put up new boxes in late fall to give the box a chance to weather a little, and to provide a roosting spot if the weather turns bitter. Birds sometimes need time to find your box, so you may not get any takers right away. If you notice that your box isn't used after the first full season, you may want to reevaluate its placement.

A good rule of thumb is to install your nesting box about 5 feet off the ground. A fairly large number of birds are comfortable at this height. Bluebirds prefer 3 to 6 feet, while chickadees, nuthatches, and wrens like boxes that are 5 to 15 feet off the ground. Again, if you know what species you are trying to attract, do a little research to determine the best height.

Like us, birds have different preferences when it comes to their neighborhoods. Bluebirds and wrens have relatively small territories, but chickadees and nuthatches want several acres of their own for raising their family. Martins are happy to share one big apartment unit (with individual compartments, of course). Some want open fields, while others prefer conifer forests or mature hardwoods. So look around your own yard and at your feeder to determine what birds you are most likely to attract, and plan accordingly. With variety and planning, your yard may be able to accommodate several boxes. They may not all be used every year, but it's nice to know they are there.

Some birds, like our little Carolina wrens, are pretty comfortable near our house. They inevitably choose our porch planters for their nests. But

most birds like some privacy, so it is best to place your box away from common traffic areas in your yard, both for the bird's sake and your own. If a bird chooses to nest near your path to the garbage can, you may find yourself attacked every time you take out the trash! An east-facing location away from the prevailing winds is ideal, and there should be adequate space for easy entry.

LEFT: A well-maintained nesting box will provide many seasons of shelter for nesting bluebirds.

Safety Considerations

Nothing is more heartbreaking than watching over a nest only to find one morning that it has been savaged by predators.

Snakes have no problem entering a nest that is placed on a climbable surface. They are perfectly comfortable with a vertical climb up a tree or a porch railing. Raccoons and cats are always on the lookout for an easy meal, too. It is a pretty simple matter for them to reach into the box and snatch the eggs, or the nestlings. The best way to help reduce the chances of a marauding mammal or snake is to mount your nesting box on a freestanding pole with a predator guard.

Squirrels are generally not interested in eating birds or their eggs, but they can cause problems by chewing on entry holes. If you experience this problem, the best solution is to line the area around the entry hole with sheet metal.

Some birds are nest predators, too. The European starling is notorious for entering birdhouses and killing both parents and nestlings. They are too big for a 1 ½-inch opening, however, so they are relatively easy to control. House sparrows, on the other hand, can be a real nuisance, and are much harder to remove. Crows, blue jays, and cowbirds may also raid a nest.

Visiting Etiquette

If you were a brand-new mother of four, would you want visitors? Probably not. Birds feel the same way. They would prefer to raise their babies in private, and will aggressively defend their nesting area from intruders. So if you can avoid a nesting area, your birds will appreciate it.

However, if you are careful and follow a few guidelines, it is okay—and maybe even desirable—to take a peek now and then. Monitoring the nest box allows you to track the species in your yard and to make sure that predators haven't caused problems. If this is a subject of interest to you, the Cornell Lab of Ornithology sponsors a program called NestWatch that anyone can join. The purpose of their program is to allow regular folks to report nesting species and behaviors to their centralized database so that they can track the reproductive health of North American birds. You can find more information on their program at www.birds.cornell.edu.

To monitor a nesting box, approach the nest and make a little noise to alert the female to your presence. A little tap on the box will allow her to fly off. Don't worry about scaring her off forever. She will return again when the coast is clear. Open the box and take a quick look. Note the color, size, and number of eggs. Don't open the box in rainy weather, and avoid it when the nestlings are about to fledge.

Maintenance of Nesting Boxes

After the baby birds have fledged, remove the nest from the box. Most birds won't reuse a nest, but will be happy to reuse a clean box. If there are unhatched eggs or dead babies, remove those. Take down nest boxes once a year in the fall and give them a good cleaning with a scrub brush and a mild bleach solution, to kill mites and other parasites left behind.

Providing Nesting Materials

My birds provide me with a steady source of cup nests. I remove the ones I find in my planters; the wrens who come every year will want to build new ones. I also pick up blown-down nests and remove nests in really inconvenient locations, like the gutters. It is fun to examine these nests. Look

ABOVE: Easy-to-access nesting material makes nest building a quicker and easier chore.

closely and you will see an amazing variety of materials, from bits of cloth and yarn to dog hair and flower petals.

Although some birds will use the same nest for a second brood, song-birds won't reuse a nest the following year. That means a lot of hunting and gathering for suitable building materials every season! We provide nesting materials in the spring. We use a suet feeder and fill it with bits and pieces of favorite building materials, like yarn scraps, bits of fabric, and soft grasses that we have cleared out of the garden. Then we hang it near the nest boxes for birds to help themselves.

I have a lot of dog hair at my house—the dogs I raise have soft undercoats that I regularly have to brush out. I stuff a separate suet feeder with the fur and hang it in a sheltered location; it's easier for the birds to find, and it stays dry. I always get a kick out of watching the birds collect the fur, and I think it's good karma for my dogs to have contributed something to the well-being of the birds. Keep in mind, though, that if your dog is treated with tick prevention or some other medication, the hair may contain toxins, so only offer it up if you are confident that it won't harm a little bird.

Making Your Yard Bird-Safe

Household Pets

In a word—no outdoor cats! Birds and cats are like oil and water, so to speak, so if you have outdoor cats, attracting birds to your yard adds a layer of stress and predation that backyard birds just don't need. If you want birds in your yard, keep only indoor cats.

Dogs are a consideration too. I have dogs, and for the most part, everyone in the fenced-in backyard knows their place. I don't place feeders or mount nesting boxes within the confines of the dog yard. My dogs don't pay much attention to adult birds, but occasional squabbles break out anyway, gener-ally because the dogs' presence can worry a cardinal or jay that have decided to nest in the lower branches of smaller trees or azalea bushes. I often walk the yard for signs of nests. If I see one, I pay careful attention to it. My biggest concern is for the fledglings. A dog can do a quick snatch-and-grab of a baby bird faster than you can imagine.

I have one darling little dog named Maya who has raised a number of beautiful litters. She is a mother extraordinaire, and will hover and fuss over every youngster she can find, canine or otherwise. One day, I saw Maya trotting quickly into the house. Because she seemed intent on something, I followed her. Sure enough, she went to her dog bed and dropped a fledgling blue jay into the center of it. Then she lay down to watch it. Being such a good mom, she just wanted to take care of it.

Of course, my heart stopped at the sight of that tiny baby so near the jaws of a supposed predator! I was heartsick. The dog's yard was obviously where it had come from, but even with the dogs back in the house, I knew I couldn't return it there. I quickly picked up the entire bed and carried it to the patio outside the fenced-in yard. It hopped off, amazingly unhurt from the short trip in Maya's mouth. She and I watched as it hopped about, exploring the patio. From another part of the yard, we heard an adult calling. Soon, the fledgling fluttered its little wings and landed in a nearby shrub, chirping loudly. Fledglings have a hard-enough time getting started in life, but it seemed that this one was back on the road to survival. Still, I had learned my lesson: It's important to be extra vigilant during nesting season.

On the other hand, your family dog can be handy when it comes to squirrel control! Our dogs routinely patrol the enclosed deck to chase away squirrels at the nearby feeders.

Windows

An enormous number of birds are killed every year by flying into windows. Experts estimate that window strikes are responsible for the loss of more birds than almost any other environmental hazard, second only to the destruction of natural habitat. The number of birds that die every year because of windows ranges up to well over 100 million, with some estimates as high as 1 billion.

Less deadly, but still frustrating, are birds that catch sight of their reflection in a window and imagine it to be a rival that they must challenge. Although jousting with an imaginary bird won't kill him, it's annoying for you, and he could spend his time peacefully conducting more-useful activities.

RIGHT: This cat is hardly welcoming to backyard birds!

It's important to understand that birds can't make sense of artificial glass and reflective surfaces, so make sure your birds are safe around your windows. Take a walk around your house and try to see what your birds see. If you have plants at your sunny windows, be aware that birds may see them and think they can reach them. Or, if you have another window visible in the background, birds may view this as a fly-through area. Sometimes a window simply reflects the landscape around it, confusing birds into believing that it is just more yard. Large plate-glass panes are particularly dangerous in this regard.

Try these tricks for camouflaging your windows:

Slow Down the Approach

At our house, we have shrubs and short trees in front of a number of the windows. Birds tend to perch there and don't usually come in with any velocity. Feeders that are very close to the windows are generally okay too. Birds that are traveling toward the feeder are slowing down rather than speeding up on descent, and thus are less likely to injure themselves on approach.

Move Feeders Farther Away

If your feeder isn't within a foot or so of the window, move it to at least 10 or 20 feet away. This may help avoid the possibility of a deadly collision.

Eliminate Inside Enticements

Remove plants or brightly colored objects that may be causing curiosity. Close your blinds or curtains, soap your windows, or add window film so

BELOW: Our dog watches for signs of squirrels and other animals that may cause problems for birds.

LEFT: Your yard may be able to accommodate several different nesting families every spring.

birds can't see inside. This won't eliminate the exterior reflective surface of the window, but it may reduce the number of birds who fly into the window.

Break Up the Reflection

On the outside of your window, hang fabric ribbons, strips of tape, or hawk silhouettes from chains to reduce the reflection. Objects that move or flutter work best.

Add Window Screens

Window screens may not entirely break up the reflection, but they will provide a cushioned surface for birds to bounce off of. If you don't have screens, there is bird netting made for windows that can be easily installed.

Consider Your Car

One day I watched a male Eastern bluebird rush from one of my car mirrors to the other, back and forth, challenging first the bird on the left, and then the one on the right. I thought the poor little guy was going to be driven to distraction with these two rivals for his territory so close to his home, so I got two small sacks and covered the mirrors. Later, I moved the car away from that parking spot during the rest of the nesting season to give him some peace of mind.

Yard Maintenance

Your Lawn

That beautiful expanse of green is attractive all right, but at what cost? Traditional lawns require a lot of water, fertilizer, and pest control to keep them at their peak condition, but many of these methods run counter to creating a healthy bird environment. To have a lawn that is healthy for wild birds, as well as for pets and children, avoid the use of pesticides and other chemical treatments. Keep the following tips in mind to maintain a bird-friendly lawn:

- **Choose the right grass:** Drought- and insect-resistant forms of grass that are well suited to your area give you a big head start when it comes to lawn care. For areas near your house where you would like a more-manicured look, choose a tough fescue or rye grass that has been recommended by your local nursery for easy care.

- **Consider going wild:** For areas farther from the house where you can tolerate a wilder look, consider letting your lawn go free—dandelions and other "weeds" actually produce seeds and attract insects that a traditional lawn wouldn't.

- **Avoid the close crop:** A simple step like cutting your lawn to a slightly longer length is good for several reasons: Longer grass conserves moisture, creates more food for the plant, encourages root growth, and establishes a better environment for insects and earthworms. Mow only when needed rather than putting yourself on a rigid schedule.

- **Mulch your clippings:** Instead of raking or hauling off the clippings, use a mulching mower that leaves tiny clippings behind. These clippings decompose to enrich the soil naturally.

- **Aerate your lawn:** Compacted soil doesn't allow root systems to function properly, so aerate surfaces that get trampled down over time. Aerating the soil creates holes that allow your soil to breathe, and make it easier for water and nutrients to reach the roots. If you have particular high-traffic areas, you might want to consider installing a stone path to eliminate the hassle altogether. This simple step can make your lawn healthier and greener all by itself.

- **Check the pH:** Lawns that are too acidic or too alkaline are inhospitable to the organisms that live in the soil. Test your lawn soil's pH level. A good lawn should have a pH of 6 to 7. To adjust the pH of your lawn, use a light top dressing of organic fertilizer.

- **Weed by hand:** It's easier to spread some killer pesticide on the lawn to eliminate all those dandelions, but it's healthier (and better exercise) to just dig up the weeds in your otherwise-immaculate lawn. A dandelion weeder is a handy tool that lets you remove these weeds at the root level.

Cleaning Up Your Yard

This is the easiest part of making a bird-friendly yard! Birds actually *like* *it* if you keep things a little messy. They don't want everything neat and orderly. While we try to manage our lot to accommodate their desires, we do like things to be relatively neat right around the house. When we were first getting started on this property, there was a small landscaped corner right near the house that we generally kept pretty tidy. One year, we just ran out of time and let that corner fall apart. The herbaceous shrubs toppled, the weeds ran rampant, and the flowers stood on their stalks. Before we knew it, a number of birds had staked it out as one of their favorite places. So we added feeders there, left the toppled shrubs, let the weeds stay, didn't rake the leaves, and unapologetically renamed the corner "Bird Central."

Dead leaves, brush piles, even your dead garden flowers—these are all very appealing as autumn sets in. So even if you prefer your yard to be orderly, consider places where you can relax a bit on the cleanup.

Pruning Shrubs

We have a lot of azaleas, and they need to be pruned every year. But these shrubs are also prime nesting areas, and hummingbirds love the flowers, so we hold off on pruning them for as long as we possibly can. Even when we do get around to the job, we trim entirely by hand to avoid upsetting birds that use the area for shelter. Consider a light touch on your shrubs, and put away the electric hedge trimmer!

Leave the Leaves

You don't like picking up leaves, and the birds don't want you to—that sounds like the makings of a beautiful arrangement! While leaves clearly need to be removed from sidewalks, driveways, and lawns, are there any places where you *could* leave them? We keep a thick mulch of leaves under our shrubs rather than manicuring the area. The decaying leaves provide nutrients for the plants and an insulating layer that retains moisture. Leaf piles provide a great area for insects to thrive. You will see ground-feeding birds foraging there regularly. We always have cardinals, towhees, and robins poking around our leaf layer.

ABOVE: If you can, allow some leaves, to remain in your yard for the birds. Leaf piles offer a great place for insects to thrive, which make for good eating.

Beautiful Winter Gardens

To a bird, a beautiful winter garden has standing stalks of old flowers full of seeds, leftover berries, and hidden insects, plus dead perennials and ornamental grasses for shelter. Clean up your flower beds a little at a time. In late summer and early fall, pull out the tender annuals that decay and turn mushy, but leave as much as you can stand into the winter. Flowers like sedum, coneflower, zinnias, marigolds, hydrangeas, and yarrow provide bugs and insects long into the cold-weather season. I clean out my flower beds in late winter, right before spring is about to break. Even in very snowy areas, standing flowers dried on their stalks are a picturesque element to the landscape and a pleasant stop-off for winter birds.

Overlook that Brush Pile

Cleaning up after a hurricane one summer, my husband started hauling branches, broken limbs, and cut logs to a sunny area near the edge of the woods. Pretty soon he had the makings of a great brush pile. We leave that area for the birds. Every year, we add windfall branches and pruned limbs to the pile. I know the average backyard may not be able to accommodate

ABOVE: Dried hydrangeas add winter interest as well as a great hiding place for winter birds.

what might seem like such an unsightly feature, but if you have a larger property, consider creating a bird-friendly brush pile rather than burning it. It makes great winter shelter, and may even be used for nesting by ground-nesting birds.

Learn to Love Dead Trees

It breaks my heart when we have to make the decision to take down a tree. Sometimes it simply has to be done for safety's sake, but I can't help thinking about all the birds that would love to recycle it for their own uses. With over eighty cavity-nesting birds in North America, one good dead tree can provide food and shelter for many seasons. We leave anything standing that isn't a danger to the house or an impediment to the road or the neighbors.

Control Insects Safely

The subject of insects is one area where humans and birds often disagree. From a bird's perspective, the more bugs the better. Unfortunately, we do not share their enthusiasm. Mosquitoes and blackflies are a particular area of disagreement. It is okay to control the insects in your yard, but it is really important to do so safely. There are products on the market that attack the larval stages of mosquitoes and blackflies. These are quite effective in preventing this population before they hatch into flying, biting machines, and they are safe for mayflies, dragonflies, and other non-nuisance insects.

To replace the commonly available insecticides and pesticides in other areas of your yard, consider organic or natural alternatives. There are a lot of products on the market. It is wise to do research on what your particular "pest" problem is, rather than buying a generic product that may literally be "overkill." Some natural products are essentially home remedies that may give you limited success. Others are more sophisticated, targeting and interrupting the breeding cycle of particular insects.

I have a particular gripe with Japanese beetles. Not only do they make a mess of my foliage, but one of the birds that finds them tasty is the European starling, a bird I would rather not attract. I set up bug traps or handpick Japanese beetles from plants instead of spraying them. But the most effective way I have found to control them is to treat the lawn with a natural product that inhibits the growth of grubs without endangering other creatures.

Skip the bug zappers, too. Those zappers kill lots of bugs, but most of the bugs they kill are not even pests anyway. The zapper may be killing a few of your mosquitoes, but it is taking out far more of the beneficial insects that your yard and your birds need.

Weed Good Deeds

Skip the Roundup if you want a healthy yard! Weed killers are not just toxic to weeds. They are also harmful to beneficial plants and any animals and humans that come into contact with them. To control weeds in your lawn, start with a good native grass that will have a better chance at withstanding attacks from disease, weeds, and fungus.

Use natural alternatives for eliminating weeds where possible. A spray of 100 percent-strength white vinegar will kill weeds between cracks in sidewalks or patios. It won't be as fast as Roundup, but it does work. Of course, pulling the weeds out by hand is part of a great exercise routine!

" Bird! Bird! "

—QUINN FEDORA, age two

7

Birds and Children

ALL CHILDREN are naturalists at heart. They love to walk through the woods and look up at the fluttering leaves; they delight at a bright flower; they are quick to make a collection of stones or feathers or sticks. And they love birds!

Your backyard birds are one of the best ways I can think of to get a child started on a path toward understanding the world around him and the impact we have on it. Children are naturally attracted to birds, and why not? Birds do pretty cool things. They fly and they hang upside down. They sing and call and quarrel and build nests. They give us a window into what it is like to live on earth without the benefit of houses and TVs and video games and shopping. They allow us the opportunity to stand in the yard and *just listen* to what the world is doing when we are out and about in our frantic life.

At two, our grandson Quinn made it part of his morning routine to run to the window and see how the birds were doing. *Bird* was one of his favorite words. It was always fascinating to watch him. Toddlers note the movement of every bird, and are quick to spot the nuthatch on the trunk of a tree or the little wren that has flown to the nearby holly bush. He liked to help fill

LEFT: Nest of baby song thrush

the feeders and put out oranges and suet. He supervised as his grandpa installed new bluebird houses and helped to collect items for the birds to make their nests.

Young children can take an active role in taking care of the birds. Pine-cone treats, simple birdhouses, and bird counts are great ways to get them involved. Nature hikes and beginning life lists hold a fascination for children who start young.

Birds and Nature: Making the Connection in the Backyard

From the moment our children were born, we took them everywhere we went outdoors. They hiked through wilderness areas, followed as we went fishing or bird-watching, and collected rocks, flower specimens, and water samples alongside their dad. At home, they watched our birds from the window and learned their names, habits, and lifestyles. They have seen eggs nestled into nests, babies breaking out of their nests, and fledgling blue jays taking their first flight. As they got older, they witnessed the shock of a hawk sweeping down to snatch a songbird and the heartbreak of a marauded nest. And they saw what happens when human lifestyles and birds' needs collide—a downed tree with nestlings in it, a bird striking a window, domestic cats wreaking havoc.

Through all of this, they have developed an appreciation for the natural world and our place in it. They understand that there is a world other than our own where creatures live and love, work hard for food, and sometimes struggle to survive. As they have gotten older, they have made the connection that what humans do really has an impact on others, and that we have a responsibility to respect and nurture those lives.

Babies and Toddlers

You are never too young to appreciate birds. Starting at a few months old, we carried our children around the yard to listen to the birds. We paused and held our finger in the air. "Listen," we said. Then we repeated what the bird said. "Cheep, cheep, cheep." Babies find this very amusing, and before we knew it, our kids were pointing in the air when they heard a bird, and looking at us expectantly. "Cheep, cheep, cheep!" always produces a belly laugh with a baby.

As toddlers get older and more mobile, they want to do everything we do. This is the perfect time to let them start helping. They can hold the hose to fill the birdbath, scatter millet on the ground, and supervise as we hang the feeders. We can take them to inspect the nesting boxes. At the feeder, we sit at the window with them and name each bird. In a backpack, they go with us as we hike through our local nature preserve, often spotting birds before we do.

The only goal at this age is to expose them to the wonders of nature. This is a fertile time for establishing birds and nature in a child's life. I am always amazed at how in tune young children are with the natural world. They notice everything the birds do, and will often be the first to point out if something different is going on at the feeder.

Young Children

As children get a little older, they can take a more active part in caring for birds. Making pinecone treats and decorating a Christmas tree for the birds is a great way to get them involved.

One of my favorite activities at this age is having children collect nest-building materials for the birds. It is fascinating to watch birds swoop down and grab a bit of ribbon or string, and we often try to guess what type of bird is going to want what type of material. Sometimes we just sit on the back deck and offer out one piece of yarn at a time to see which bird will come along to select it.

At this age, children love stories about the different birds in their yard. We tell stories of the rascally jay and his family, or how the titmouse family is staying together for the winter, or how the juncos have left their mountain home to come into town for the winter. We wonder where the flocks of red-winged blackbirds are headed, and pull out their range map to choose an area we'd like them to go. We daydream about leaving with the humming-birds to sunny South America. Sharing the life stories of birds is a great way to expand a child's sense of the world and gets them interested in geography and mapping.

It is also a good lesson about weather; understanding how hard it is to live out in the cold and the snow gives children an appreciation for what birds have to do to survive in bad weather. Watching the birds puff themselves up in the cold is a great lesson on the power of down feathers. When our children were small, they were so taken with this particular bird skill that they wrapped themselves in down comforters and sat in a snowbank to see if it really worked.

Outside of your yard, children can start to make their own sightings. A local park or nature preserve is the ideal place to get started. Field guides made specifically for children help them begin to learn identification skills. Binoculars are a big hit at this age, and there are some on the market that are well suited for little hands.

Children are fascinated by feathers. Examining and identifying feathers together can be a fun activity at this age, and your child may want to start a collection. Just make sure they wash their hands after handling feathers!

School-Age Children

Many school curriculums support bird-watching as a great way to teach ecology and conservation. Lessons on habitat and anatomy are often integrated into science lessons. Grade school–age children can begin to keep their own journal to note their sightings. A sturdy, child-safe digital camera is fun for them at this age. You won't necessarily get great pictures, but the idea of "capturing" a bird to take home is very exciting.

Audubon's Christmas Bird Count, an annual project that has been repeated for over one hundred years, is a great way to get involved. Local groups get together to count the types and quantities of birds in a certain area then report their results back to the national organization which tabulates the results. To join, find a local bird watching group in your area or go to www.audubon.org for more information.

Cornell University's NestWatch is another way to get involved. People from all around the country collect and submit records of nesting birds in their area. This project teaches about bird-breeding biology and engages children in a better understanding of the world around them. And what is more thrilling than getting a peek at baby birds

LEFT: Being near a nesting wild bird is a magical moment for this little girl.

in their nest? For more information on how to participate in NestWatch, go to www.nestwatch.org.

The American Birding Association has a Young Birders Group that hosts special events, including a Young Birders' Conference for thirteen- to eighteen-year-olds, held every summer.

ACTIVITIES TO DO WITH CHILDREN

Nesting Activities

Making a Gourd Birdhouse

It is fun to provide nesting sites for birds in your yard, but wooden nesting boxes are only one way to do this. Long ago, Native Americans started making birdhouses out of gourds to attract purple martins. The simplest of all birdhouses, children can produce this classic little house for cavity nesting birds with a little adult help.

To make a gourd house, you will need a nice-sized gourd. I like one at least 8 inches across. It should have a long neck. If you grow your own gourds, leave them to dry in a cool place. Once your gourd is dry, cut a 1-inch circular opening. The easiest way to create a clean hole is with an electric drill, but a pumpkin-carving knife may work too. Add a perch by drilling a small hole beneath the entry and gluing in a 2-inch dowel. At this point, the gourd can be decorated on the outside with acrylic paint. To preserve the painting, coat the gourd with a couple of coats of sealant.

Tie a rope around the neck of the gourd and find an appropriate place to hang it. You and your child may want to research what types of birds in your area may be attracted to this type of nest. For martins, you may hang it near your house, but other birds may prefer it in a more-private location.

Making a Bird Shelf

Cavity-dwelling birds get a lot of attention when it comes to housing, but robins, doves, wrens, and many other birds also enjoy having a safe, handy option for their nest-building site. A bird shelf can be an attractive option for birds that build their nests in the open, and that might otherwise choose your house gutter or some other inconvenient location. A bird shelf also makes it easier to watch your birds at their nest. If you like woodworking projects, this is a really easy one that you can do with your kids.

Use untreated lumber to create a simple 9x9-inch L-shaped shelf, and leave it unpainted so that it can weather. Then just attach it securely with wood screws under the eaves of your house or outbuilding, or in a tree that will provide some privacy.

Birdie Scrap Bag

Birds use a lot of energy to collect nesting materials, and our children like to help them find the things that they need. I always keep out a suet cage filled with dog fur for the birds' use, so the kids concentrate on finding an assortment of other types of materials. We give them each a mesh produce sack, the kind that onions come in. They collect string, yarn, and some fabric ribbons and cut them into short lengths. They take hair from combs and brushes around the house, and from the dogs, and select dry grasses and small twigs. Some people use dryer lint, but we avoid it because of the chemicals it contains.

We enlarge a few of the openings in the mesh bag and hang it near the feeders so birds will notice. It is fun to watch and see who comes to visit and select something. After nesting season, we collect nest cups and look to see if any of our materials were used.

Bird-Feeding Activities

I often have kids help me in the baking of any bird treats I make, but I always reserve the making of cups and cones for when the kids are at home. We decorate our chosen pine tree with a combination of cups, cones, fruit stacks, and commercial seed bells for the winter season.

Treat Cups

These treats are easy to make and are fun to place in the branches of our fruit trees. Let your child choose hollowed-out orange halves, coconut halves, or even pomegranate halves with a few seeds still clinging to the inside. Just push a wire through the cup (you may need a hammer and nail for the coconut shell). Fashion a hook at the bottom to keep the cup on the wire. Then curve the top of the wire to hang it from the branch. Pack the cup with a suet mixture and sprinkle sunflower seeds on top.

Christmas Bird Cones

We always make these to decorate the trees outside during the holidays. Collect nice-sized pinecones or buy them at your local craft shop. Pinecones with wide gaps between the scales are perfect. Pack the cones with peanut butter or your favorite suet mixture, pushing deep into the crevices. Then roll each cone in a sunflower-seed or crushed-peanut mixture. Tie a red ribbon at the top and hang on your tree.

Fruit Stacks

These are easy to make, and it's fun to see who comes for the fruit. Take a coat hanger and straighten it out, using wire pliers to make a knot at the bottom to hold the fruit. String quartered apples and oranges and fashion a hook at the top. Hang in an open tree. We sometimes alternate peanut butter—covered apple slices with the plain ones.

RIGHT: A Carolina wren helps itself to a holiday pinecone treat.

> **"** I soon discovered that the most satisfactory outlet for expressing my excitement over birds was the camera, rather than either pencil or brush. **"**
>
> **—ELIOT PORTER**

⑧

Bird Photography

IF YOU have an interest in photography, it's only a matter of time before you start thinking about how much fun it would be to capture your favorite subjects! Photographing birds takes many forms. Perhaps you just want to capture a specimen for later study at home, or maybe you want to document a rarity to share with others. Advances in photographic equipment, like high-speed digital cameras and lightweight spotting scopes, have made it possible to achieve almost-instant results. One can leave the field, download the file, and share the news of your find with others in just minutes.

For many, photographing birds is a fascinating hobby that adds tremendously to their experience of watching birds. Bird photography gives one the opportunity to really live with the birds for a short time, to engage in simple observation of their behaviors and activities. If you are sociable by nature and enjoy sharing your hobby with others, there are a number of guided field tours that take you to prime birding locations. You may have seen photographs of photographers lined up side by side, telephoto lenses at the ready, prepared to document major migrations or exotic sightings. But I think that at its best, bird photography is a quiet and solitary activity. My

LEFT: Dove

favorite moments are sitting in the backyard, waiting to catch the perfect portrait of my favorite titmouse, or sitting completely still near the berry bushes at the local nature preserve to see who comes to feed.

Getting started in bird photography requires a few basics. You need to select the right equipment and develop some skill in using it. It helps to have a good tripod and a strong, steady arm. Oh, and most of all, you need to bring along a lot of patience!

The Right Equipment

When it comes to getting started in bird photography, people always ask, "How much camera do I need?" The answer to that question really depends on what kind of results you wish to achieve. Advances in digital photography have increased your options tremendously since the days of film cameras. To record birds for your files, a high-quality point-and-shoot may be a perfect addition to your birding kit. But if you want to get serious about bird photography, a camera with interchangeable lenses is still the way to go.

Point-and-Shoot Digital Cameras

We never go anywhere without our digital point-and-shoot camera. The latest technology allows us to carry a high-pixel-count camera with a good zoom for an investment of less than $300, and a weight of half a pound. Ours is always in a pocket for quickly recording birds and snapping pictures of family and friends. In the backyard, where you may have good luck luring birds close to you, a point-and-shoot camera can produce images you can be really proud of. Along with your camera, buy a couple of 4GB memory cards so you don't run out of storage space when you're out in the field. Choose a digital point-and-shoot camera with the following features:

Resolution: 10 to 12 megapixels
Zoom: Digital zoom of 4X (You will also see a designation for the optical zoom rating; this number rises to 12X, and may be okay for using to catch a specimen for closer study, but image quality is generally poor.)
Focus: Auto focus
Exposure settings: Choice of manual and automatic
Battery: Rechargeable

RIGHT: A cooperative bird can be captured satisfactorily with a good point-and-shoot camera.

RIGHT: A cooperative bird can be captured satisfactorily with a good point-and-shoot camera.

Digital SLR Camera

While just about any camera can get you started, if you want to capture sharp, bright, properly focused bird images, choose a digital single-lens reflex camera (SLR). Technology is advancing very rapidly in this field, and there is an amazing selection of excellent-quality digital SLR cameras available in almost any price range.

The Camera Body

There is so much to say on the subject of digital cameras that it would take an entire book to tell everything. Like all things electronic, digital camera bodies offer more features than most people will ever use, but there are a few basic factors that make certain cameras preferable for the birding photographer:

Pixel count: A simple way to decide on what pixel count you need is to ask yourself, "How big a print might I want to make?" A 6MP image can produce a decent 8x10 image, but for an 11x14 image, you will need 10MP. But if you are shooting birds, chances are you will often crop your image. That's why we go for as many megapixels as we can get.

Sensor: The sensor is a complicated thing, but it helps to think of it in the same way you once thought of film size. You can think of a full-frame sensor as a 35mm film size, roughly 36mm x 24mm. Smaller sensors, however, can be very useful for bird and wildlife photography, because they require a smaller lens to provide the same angle of view and zoom length. A smaller sensor will give you a 1.5x–1.6x magnification advantage, and save you the weight of carrying a bigger lens. For that reason, our bird camera uses a sensor size of 22.3mm x 14.9mm.

Sensitivity: Sensitivity is a measure that is roughly equivalent to film speed. Standard digital cameras assume an ISO of 100 to 200. The sensitivity setting lets you tell your camera what ISO to shoot at. The more light coming into your sensor, the faster the shutter speed you can set. This feature lets you "simulate" faster film, allowing you to capture low-light images, but without the reduction in image quality that these fast films once produced. Choose a camera that will allow a wide range of ISO choices. While we like to have the ISO set between 200–400 for best lighting flexibility and sharpness, we often push to as much as 1600–2000 to catch a bird in motion, or for lower light conditions.

Modes: Another basic feature that we like in an SLR camera is the ability to use a full array of manual options, both in exposure and focusing. While there are many times when you can keep your camera on fully automatic mode and get good results, there are often situations where you need to use manual focus and exposure to ensure a sharp, properly exposed shot. A good digital SLR will allow you to move between auto and manual instantly as needed.

Lenses

The primary reason for choosing an SLR camera is to take advantage of the ability to use interchangeable lenses. You will need a good telephoto lens to get consistently good bird photos. Consider buying a minimum focal length of 400mm if you plan to shoot in the field. Serious photographers use even longer lenses, from 600 to 1000mm is commonly used in bird photography.

The other lens we often use in field photography is a zoom lens. For general bird photography, we have a 70–300mm zoom which we use often while out for an afternoon of birding. Using a zoom lens provides both advantages and disadvantages. The advantages include greater flexibility in framing your shots, less time wasted switching lenses, and less likelihood of getting dust or dirt inside the camera. The disadvantages include heavier weight, a maximum focal length that may not capture distant birds, and higher cost for quality optics.

After you have selected the length of your lens, it's time to consider aperture. Lenses are referred to as fast or slow. Fast lenses have a wide aperture, which allows more light to pass through the lens. F-stop is the ratio between width and focal length—but don't worry about the math. Just remember: low f-stop = fast lens. For example, a lens with a minimum aperture of f2.8 is fast. It allows more light into the camera than a lens with a minimum f4 aperture. Likewise, f4 allows in more light than f5.6.

While it may seem appealing to buy the fastest lens you can afford, there are a couple of serious trade-offs to consider. First, the faster the lens,

ABOVE: The ability to manually set for exposure and focus allows you to catch subtle colors and lighting. This little Carolina wren is singing his heart out in a spot of sunshine in the midst of a cloudy forest.

the higher the cost. Second, lens weight increases significantly with faster lenses. That is a serious consideration when thinking about field work.

An important recent development in lens quality has made a huge difference to long-lens photographers. This feature is called *image stabilization* (IS). IS lenses contain optical elements that correct for camera shake and stabilize the image to give you better sharpness than you might normally get with hand-holding a long lens. IS lenses let you hand-hold at a shutter speed of two to three stops slower than the minimum normally required for sharp images without this feature.

Accessories

Teleconverter: This is a short lens that couples between your camera and your lens to increase your focal length. Adding a 1.4x teleconverter to your 400mm lens makes it equivalent to a 560mm lens. You will only lose one stop of light by adding the 1.4x teleconverter. I don't recommend 2.0x teleconverters. Not only will you lose two stops of light, but unless your optics are of the very highest quality, your image quality also drops significantly.

Tripod: Even with a lighter-weight system and a lens with image stabilization, you will still need a tripod that will support that 400–500mm lens. We virtually never leave home without our tripods. As novice photographers, the best advice we ever received from a pro whose work we admired was, "Buy the best tripod you can afford and use it all the time." We took his advice, and our photos improved immediately. We own lightweight carbon-fiber tripods. These can be pricey, but there is a wide range of models on the market, and many reasonably priced carbon-fiber and aluminum tripods available. You will also need a good ball head and a quick-release bracket that allows you to quickly and easily mount and dismount your camera from the tripod head.

Filters: Buy a good circular polarizing filter for each of your lenses to cut down glare on bright days and to photograph around water. Keep a skylight filter (or 81A or B filter) on your lens at all times to keep it from getting scratched and collecting dust.

Field/storage bag: A good comfortable field bag is a must. It should be highly water-resistant for those drizzly days when the birding is good but the conditions aren't.

Flash: A good flash system is great, but for outdoor bird photography, it gets complicated. If you are interested in serious field work, research portable lighting systems and read up on techniques for making these systems work well in birding conditions.

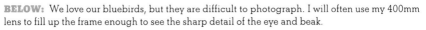

BELOW: We love our bluebirds, but they are difficult to photograph. I will often use my 400mm lens to fill up the frame enough to see the sharp detail of the eye and beak.

ABOVE: In a shaded area, image stabilization means the difference between a missed shot and a great one.

Lighting

If you are going to be up with the birds, early morning is where it's at for photography. Birds are busy in the morning, and if you can get yourself out there with them, you may be rewarded with beautiful lighting.

Full-sun conditions can be great for capturing the vibrancy of brilliant plumage and illuminating darkened forests, but bright sun can also produce unacceptable glare on reflective surfaces, and high contrast and backlighting when the subject is facing the wrong direction. Full sun can also produce strong shadows that play havoc with your composition. Filtered sun or flat lighting conditions like overcast days are often more conducive to good photography.

While it might seem easier to photograph in the middle of a bright, sunny day, more often than not you will be disappointed with the final images you download to your computer. Bright sun produces high-contrast photos with deep shadows and areas of light that have blown out all color. Reflections and bounced light can ruin the shot, and colors will be less saturated.

Make the effort to photograph early and late in the day. Birds are more active in the early-morning hours and often late in the day as well. Get out on overcast and even drizzly days. This can often be the best light

LEFT: I had a pretty good idea where I would find this pair of downy woodpeckers, so I sat down with my tripod to wait for them to give me a good shot.

because it is even. There are fewer shadows, less reflective light, and the colors will be much more saturated. Make specific trial shoots in these different conditions in the same place and see for yourself.

Birds are often very active before, during, and after storms. This year's surprise 28-inch snowstorm is a good example. Needless to say, here in Virginia, both we and the birds were ill prepared for it. The birds in our yard never stopped feeding throughout the day, and we kept going out to clear off the feeders and replace food for the ground foragers. But as the snow subsided and the sun came out, the photos were glorious.

Bottom line: To dramatically increase your success with photographing birds, get to know your birds, their habitat, and the environmental conditions where you can best capture them. And then get out there!

Capturing Motion

When we are photographing birds, motion is pretty much a constant state. So even if you are photographing at the feeder, you have to be prepared to stop motion. As with any type of wildlife photography, concentrate on sharp eye focus. Soft focus anywhere else in the shot may be acceptable as long as the eye is in focus.

If using manual exposure controls, we like to keep the f-stop in the f9–f11 range. This gives us a little depth of field even with the longer tele- photo lenses. It makes it much easier to ensure we get the eye in sharp focus, even in the lower light ranges of early morning or on overcast days.

To stop bird movement, your shutter speed should be at least 250th of a second. Shutter speeds of 400th–500th of a second and higher are even better.

Backyard Birds

Before you pick up your camera to go out after birds, it really helps to know the habits of your subjects. If you already have a backyard feeding program, this is the best place to begin. Chances are you already know something about the habits of the species who visit regularly. Some of your feeder birds may already be accustomed to your presence, and will have no problem seeing you nearby. Proximity is everything when it

The Serious Birder's Photo Outfit

Digital SLR camera body with the following features:
Pixel count of 12 to 18 megapixels
Sensor 22.3mm x 14.9mm CMOS
Sensitivity: ISO range of 100–6400
Exposure Modes: Manual and adjustable
Focus: Auto and Manual
Lenses
100–400mm f5.6 zoom lens with image stabilization and auto focus
500mm f4 lens with image stabilization and auto focus
Accessories
1.4x teleconverter
Polarizing and UV filters
Water-resistant gear bag
Lightweight carbon-fiber tripod and quick-release bracket.

Bright, sunny conditions require careful metering and the right filter to get a sharp shot. My brother caught this osprey in the midst of its morning hunt.

comes to getting a really good bird photo, so it helps to set up feeders where you can get a ringside seat.

In our backyard, we generally use the "sit and wait" method of photography. Using food or water to bring the birds where we want them, we then position ourselves to shoot what happens in that area. We have a couple of feeding setups that we use specifically for taking pictures of our birds. One is right near the house. We have a window that is easy to open where we can set up a tripod and wait in comfort. We open the window wide and pull the curtains closed, leaving only the telephoto lens exposed. We have strategically placed feeders in trees with nearby branches that can be used to frame a good shot. The best feeder shots don't show the feeder at all, so it helps if there are easy-to-shoot areas where birds alight when not at the feeder.

For our ground-feeding birds, like cardinals, juncos, and wrens, the nearby boxwood and hydrangea bushes provide great picturesque backgrounds, and we know that we will often see them sitting comfortably there, waiting for their next turn at the feeding station. We also add pine branches and logs to the ground-feeding area to achieve a more-interesting setting than the ordinary "bird pecking at birdseed" shot.

Our nuthatches and woodpeckers are very predictable. It is easy to set up shots on the tree trunk where they always go when leaving the feeder. They are relatively calm birds, too, so they make an interesting and engaging subject.

For the birds that tend to feed and flee, like our titmice and chickadees, we are more reliant on shooting them right at the feeder. These birds tend to grab one seed and run off with it, often disappearing completely with their loot. As they become more comfortable with the feeder, though, they may hang out long enough to give you a good shot, and even stay in a nearby branch to eat their seed.

Finches, of course, are always having a food party, and will stay to feed longer, providing ample opportunity to take their picture. Everyone has the ubiquitous tube-feeder shot of a flock enjoying their seed

LEFT: Soft and dreamy, this red-wing blackbird poses in the snow storm.

ABOVE: This red-tailed hawk looks magestic in flight.

together. But these birds are great for practicing tight, close-up shots. To get more-natural finch photos, I have tried a tea-stained finch sock mounted to a stump, as well as a platform feeder camouflaged with branches. These provide a better background than the standard acrylic tube.

We build two feeders especially for photographing birds. The first is a simple log suet feeder. We hang this one away from the other feeders in a picturesque birch tree. It provides a natural setting for woodpeckers, nuthatches, and other trunk-clinging birds. To make a log feeder, select a log with a diameter of about 6 to 8 inches. Drill two 1-inch holes opposite each other on the log and pack with suet. Add an eye-hook screw at the top of the log for hanging. Hang the log with the holes to the sides so the suet isn't visible to your camera angle. (Note: This feeder is extremely popular with our squirrels, which will chew on it and do their best to knock it down, so when we aren't taking pictures, we take it down and put a plain wire-cage suet feeder in its place.)

The second feeder location we have is a large log that we use as an all-purpose ground feeder. It is situated well away from the house near the edge of the woods, with easy access to bird hiding places and escape routes. We hollowed out the top to create a sort of trough to hold food. We

also scatter seed and cracked corn on the ground around the log. We have an easy view of it from a patio, and nearby trees allow us to watch without disturbing the activity. It has attracted the typical ground-feeding birds, as well as wild turkeys, deer, raccoons, squirrels, and chipmunks. Some people say that mammals are a nuisance and chase away birds, but we find that they all use this feeder without any major squabbles. This type of feeder may not be for everyone; besides attracting quite a variety of visitors, it also requires quite a lot of food, particularly in the winter!

Once you have mastered your feeder birds, look around your yard to see where your birds go when they are not at the feeder. Holly bushes, dogwoods, and Virginia creeper berries provide natural forage, even in the dead of winter, and can be a great place to set up your camera. We have one area of dense shrubbery that seems to be especially appealing to our birds, and we sometimes set up the camera to focus there for an entire day.

Water is another great place to take bird photos. Your birdbath may be an ideal location, especially for birds that have no interest in your feeders. We have one bath that we shoot at often. It is right under a dogwood tree, which provides some security for the birds, and gives them an easy branch on which to sit as they finish their drying and preening process. A fountain or spray attachment that creates running-water sounds or creates a spray is a great attractor too, especially on a warm day.

RIGHT: Even the ordinary bird-pecking-seed shot can make a sweet photo, as this cooperative song sparrow illustrates.

RIGHT: When photographing backyard feeders, we often keep the camera trained on nearby branches. This tufted titmouse flew from our feeder to a nearby tree.

Photography in the Field

While "sit and wait" photography works pretty well in the backyard, where your birds' habits are fairly well known, it is often necessary to stalk birds in the field. Birds in the backyard or even in local parks may be habituated to the presence of people, but like any wildlife, they feel threatened by the direct approach. Your goal is to get as close as you possibly can. The bird's goal is to stay out of harm's way. With that in mind, never walk directly toward the bird you are trying to approach. Take an indirect path, focusing your attention away from the bird. Never stare directly at the bird. Move toward the bird a little at a time, stopping anytime you notice signs of nervousness. It helps if you have done a little pre-work checking your lighting and other settings. Place your camera on its tripod or shoulder stock before you start the approach so you can set up quickly when you get as close as you can.

If you know the location of an area highly likely to produce results, you may want to make use of a blind. Setting up a blind gives you complete coverage and allows you to stay in an area for longer periods without causing alarm to the population. We carry a simple camouflage blind that is nothing more than a folding chair with a tent over it. We can put it up and

ABOVE: A congenial bunch, these American goldfinches will perch for long periods at the feeder. On this day, they host a visiting dark-eyed junco in the midst of a snowstorm.

take it down in moments. Some people find their car makes a good blind for birding areas accessible by road.

Wear drab colors out in the field. Skip bright colors and white. Camouflage clothing is probably unnecessary, since the birds may notice your sound before anything else. Therefore, choose "quiet" clothing that doesn't rustle when you walk, and lightweight shoes that allow you a degree of stealth.

Some bird species are naturally very curious and may actually come over to see what you are doing. On a vacation out West, we spent one really fun afternoon with a black-billed magpie that thought *he* was the one conducting the photo shoot. He dropped out of his tree repeatedly to check out the camera, perching on the telephoto lens and peering into the glass. He was completely comfortable with us and stayed for lunch, helping himself to bits of bread on the picnic table.

RIGHT: This friendly magpie held still for a photo shoot.

Respect for Your Subjects

One of the most rewarding times to photograph birds is in the spring. Their plumage is at its brightest, and their nesting and feeding activities make it easy to predict their behavior. But it's also a time that calls for a lot of caution and respect. Take note of any time your proximity to a nest causes anxiety. Anxious calls, a nervous mother hopping around nearby branches, and dive-bombing are all signs that it's time for you to back off.

If you want to photograph the parents building or tending to their nest, keep a good distance away and set up a blind, or provide good camouflage so that you don't cause them to change

RIGHT: Our "go-to" bush for all kinds of birds, this bittersweet vine hosts a female cardinal. We often find cardinals in the vicinity of winter berries.

their mind about the location. A nesting box or natural cavity can be easier to locate than a cup nest, and will provide a lot of activity for you to photograph while the adults go back and forth.

To approach the nest to photograph nestlings, try to wait until the mother leaves, or gently flush her from the nest, and never keep the parents away from the nest for more than twenty minutes. It is an old wives' tale that birds won't return to a nest that has been visited by a human. Birds don't have a sense of smell, so your presence at the nest won't leave behind any telltale markers for them. However, your presence can signal predatory animals about your find, so be very aware of the signs you are leaving. You can easily interfere in the process of raising their young by being in the wrong place at the wrong time. The last thing you want is to be responsible for the death of your little subjects. Be especially careful as the nestlings approach the time to fledge. In most songbird species, this is just twelve to fourteen days after hatching. A disturbance could cause them to fledge too early, a potential disaster for little ones not ready to be out on their own.

BELOW: Tiny crow nestlings await their mother's return.

ABOVE: This flock of geese offers a dramatic photo opportunity.

Outside of mating season, you can find birds where the food is. Fruit shrubs, great bug-hunting woods, dead trees, and open meadows are just some of the places that provide opportunities for good photography. Migration is a fascinating time for photography because it often brings both large, impressive flocks and a variety of species that you wouldn't see during the rest of the year.

⑨

Creating Your Own Life List

SOONER OR later, you are going to ask yourself just how many birds you have seen. It is inevitable. Birders are inveterate list makers. It may start simply, with a question like "How many different birds are at my feeder?" For me, it turned into "How many birds are at the local preserve?" Then I started wondering how many shorebirds I knew, and how many subtropical birds I'd seen. Before I knew it, I was hooked!

There is no single correct way to create your life list. For some people, making a check mark in their field guide is enough. Others have spreadsheets with location, date, and count. For some, organizing by orders and families is the only way to go. Online, there is helpful software, as well as Web sites and blogs that allow you to share your list with others. I still have my trusty journals—piles of them have accumulated over the years. Though I admire some of the efficient computer-based programs for listing, I haven't been able to get myself to automate this particular hobby. This is the

LEFT: Female ruby-throated humming bird

one activity where, for me, a soft leather-bound notebook is still the perfect accessory.

However you start, you will find it rewarding and educational to "collect" your lifetime species count. There are really only two rules when it comes to making lists. Rule #1: Only write down identifications you are certain of. If you can't verify a bird, it doesn't count, and you are only cheating yourself! Rule #2: Don't forget to stop and enjoy the bird. Some people get so wrapped up in "the next bird" that they forget to enjoy the experience.

Your backyard and neighborhood can garner you as many as 100 birds for your list. Traveling to your local parks and around your state can add another 200. Then travel to the special places around the country where birds congregate in the millions. With over 850 birds in North America, and 9,000 species worldwide, the sky is literally the limit!

Old Friends

How many birds do you already know? Pick up a field guide and start browsing. You may be surprised by just how many birds you can already easily identify. But be careful; an age-old question quickly comes to mind: If I only *think* I remember that bird, did it really exist?

Some people who start keeping lists want to start fresh. They are only willing to add birds that they see at the time they start their list. Others can recount dozens from memory, and can quickly create a great starting list. Personally, I keep a list of birds I know and track whether I see them again. I get a kick out of noting species in places or seasons I haven't seen them before. I like to track how many I saw and what they were doing at the time.

I also ask myself whether I just know that bird, or whether I have actually ever laid eyes on it. Does it count if I have heard the yellow-billed cuckoo all my life? Have I ever actually seen it? Did I just see a picture, or did I find it with my own eyes? Does it count if the bird was not in the wild? If someone else in my party saw it, does that count? These are questions for you to decide. Set your own criteria and then go for it!

Backyard Guests

One list that is an absolute must is the "guest book." We keep ours by the window of our closest feeders for jotting down who we see and when. I often

make an entry in the morning, and note other birds during the day if I see
something unusual. You will quickly develop a list of your "regulars," and
might want to note how many you see at the feeder at any given time. Because you get a lot of the same birds day after day, you may view this exercise
as redundant, but it's always fun to look back and note the day there was a
change, when someone new came to the feeder, or when a certain bird was
absent. I usually take ten minutes and write down the birds I can see together at the feeder at one time. This note from my own guest book is typical
of a winter morning:

> 2/14/08, 7:30 AM—*Cold and overcast, 17 degrees. Busy morning—*
> *they seem to love the new suet. The oranges are missing—someone*
> *made off with them during the night.*
> *Tufted titmouse, 4*
> *Carolina chickadee is back—wonder where he has been?*
> *Cardinals: 2 females, 1 male*
> *Dark-eyed juncos, 9*
> *White-throated sparrow, 2*
> *Carolina wren, 2*
> *Mourning doves, 4*
> *Red-bellied woodpecker*
> *White-breasted nuthatch*
> *Goldfinches, 7*
> *Downy woodpecker*
> *Blue jays, 2*
> *Caught a glimpse of a new bird at the nyjer seed—a grosbeak?*
> *Watch for him!*

Later in the day, another entry read:

> 3:30 PM   *Bluebirds on the fence post!*
> *Pileated flew through the backyard.*
> *4 robins in the front yard.*

If you enjoy tracking your backyard birds, you may want to get involved
in the Great Backyard Bird Count—a four-day annual event sponsored by
Audubon and the Cornell Lab of Ornithology. Over 11 million birds are
counted and recorded every year, giving scientists valuable insight into

populations and behaviors across the country. You can participate by viewing and recording your own birds for as little as fifteen minutes; then, go online and share your information. Visit www.audubon.org for more information.

Out and About

I love the added dimension that my awareness of birds brings to all of my everyday routines. I stop to listen to the mockingbird that is often in a tree near the grocery store, and I often spot the same hawk that hunts near the highway. I love to see the blackbirds in the fields out in the country, and I always pause to watch the Canada geese raising their young on a small pond on the way to my favorite garden shop.

We are very fortunate to have a remarkably good nature preserve near our house. We often refer to it as a birdy fantasyland, because it seems like there is something there for any bird that happens to be traveling through Virginia. We see eastern towhees and cardinals in the brambles, blackbirds and meadowlarks in the fields, and over twenty waterbirds near the river, not to mention seven different woodpeckers and every manner of finch and warbler. In fact, over 200 species have been spotted there. Of course, I keep a list of all the birds I see there, and my husband photographs there regularly.

A walk through your neighborhood or local park can yield additional species. Fruit brambles or an inviting garden bed may attract birds that don't come to your yard. Take nature walks near rivers or lakes to find waterbirds. Marshland is often a rich source of new birds.

To get the most out of your local bird scene, join a local birding club to connect with other birders who can show you the hot spots. Birding with local experts can open windows to birding opportunities you may have missed on your own. And it's fun to share the excitement of an unusual sighting with like-minded souls.

Farther Afield

You don't have to plan a bird-watching trip in order to see birds during your travels. My husband and I often take fishing trips that result in catching more birds for our list than actual fish. One of our greatest vacation moments was when we spotted a mangrove cuckoo while bonefishing in the

LEFT: A great egret stalks fish at a man-made pool in a state park.

Bahamas. My husband caught a glimpse of him just as he made a cast, and we both watched, spellbound, as the bird circled and landed in a tree on a nearby island. Fish were forgotten as we reveled in this most unexpected surprise. Likewise, trips to the city can often yield surprising finds. A walk through Central Park early in the morning can yield a dozen different bird spottings. Anytime we plan a trip, we do a quick search for birding areas where we are going. The most delightful and unexpected moments can happen during an hour stolen for a solitary walk in an unfamiliar park or preserve.

National parks make great birding destinations; it can be rewarding to dig a little deeper and research some of the spectacular network of state parks and wildlife refuges in the country. These areas are managed specifically for natural habitats and are almost always worth investigating, no matter the state where they are located.

If you have a hankering for real adventure, there are many bird-watching vacations available, led by experts in the field. You can plan a trip to take advantage of a big migration or one that coincides with a birding festival. Or, if plans take you to a particularly interesting location for birding, research available day trips in the area.

Luring Birds

Have you ever been out in the field and found yourself wanting to say to a bird, "Hey—excuse me! Could you just come over here a second? I'd like to see your field markings!"

ABOVE: A walk in the park during a business trip to Seattle yielded me a chestnut-backed chickadee. Another entry on my life list!

Well, actually you can, with a technique that birders call *pishing*. This is the sound we make to bring curious or reluctant birds out of their hiding places so we can get a closer look. As a technique, pishing is not very sophisticated. To pish, simply repeat a *pish-pish-pish* sound with your jaw slightly clenched. Curious birds hear this sound and come over for a closer look. It works very well with songbirds, although waterbirds seem less impressed.

Many people believe that birds may identify this sound as a scolding chickadee or an alarmed bird, or maybe even a possible insect. But I often use a similar sound to calm dogs and horses, and even a belligerent toddler may settle when they hear it. So I prefer to think of it as a universal language that means, "Hey—attention over here!"

Even the novice pisher can have good luck on his first try. Different birders have slightly different techniques, and you may find that you enjoy developing a repertoire of your own. Try to make a chattering or a clicking sound, or develop a squeak. One good way to squeak is to quickly kiss the back of your hand four times in rapid succession. This may simulate something that sounds like a scolding bird.

Because the sound you are simulating may be interpreted by the bird as a possible warning, alarm, or scolding sound, be sensitive to raising an alarm in the bird population. Most birds are more sophisticated than we give them credit for. Chances are they will pop in for a look, see a crazy human, and head back to their own business. But nesting or migrating birds that are already stressed do not need this kind of extra stimulation, so keep pishing to a minimum under these conditions.

The other technique birders sometimes use to lure birds is the playing of recorded birdsongs. With the advent of the iPhone and other handheld digital field guides, it has become easy and convenient to attract birds with a quick song of their own. While pishing tends to just stimulate a little curiosity in birds, recorded calls that sound just like them can be really upsetting, especially if they are defending a feeding or nesting area. Be aware that the sound of another bird in his territory can be very distressing, and the bird you attract may exhaust himself looking for his rival. Use these recordings in moderation, and never replay a song more than twice.

The first time I tried using an iPhone recording on an eastern towhee, I upset him so badly I immediately wished I could take it back. It was the dead

LEFT: An intimate view of the normally sociable American flamingo in Everglades National Park.

RIGHT: During breeding season, the bills and legs of the American white pelican turn bright orange. Never a solitary bird, they stay together in large colonies for nesting.

of winter and he was in a thick covert of brambles. He clearly thought I was after his berries, and came out to fight. I apologized profusely and backed away, but I noted that it took him several minutes to calm down and get back to his day. Now I use my iBird recordings strictly to educate myself about different birdcalls. In the field, I stick to pishing.

Field Gear and Accessories

The great thing about birding is that all it really takes is an interest in birds and a walk in the woods. The other great thing about birding is that, if you choose to, it gives you a chance to indulge in all sorts of great optics and gadgets! The basics are simple: A pair of binoculars and a field guide is all you really need. But even these simple components require a little research. There are a lot of choices out there.

Binoculars

With all this close examination of quickly moving objects, it helps to have a second set of eyes—the optical kind. Binoculars come in all sorts of sizes, magnifications, and size ratings. Shopping can be a bewildering experience. Choose a reputable store with knowledgeable salespeople who understand birding and can help you wade through the technical details. When I go shopping for binoculars, I go to the store armed with the answer to these three basics questions:

Purpose

What do I want my binoculars to achieve? Some people are all about **magnification power.** That's the first number in the binocular description. They want the highest magnification they can get to really bring the birds into close view. That's fine. But high power means heavier weight and a smaller field

LEFT: This eastern towhee was drawn to the recording on my iPhone. Although it turned out I was not competing for his berries, he was still none too pleased to see me.

of view. It is harder to hold a big pair of binoculars steady. That higher magnification also makes it more difficult to get a steady, well-focused bead on that bird. If you really want magnification of 10x or more, you may want to consider a spotting scope.

The second number in the binocular description is the designation of **lens diameter.** A wider lens diameter lets in more light than a small diameter. Most standard binoculars are in the range of 30–50mm in diameter. Compact binoculars are less than 30mm, and you will quickly notice a corresponding lack of brightness in these.

My husband is often behind a long telephoto lens, which means I got to pick out our latest pair of binoculars. I prefer a pair of binoculars that are relatively light in weight, have easy focusing capability, and are pretty bright. The model I chose needed to fit comfortably with relatively thick eyeglasses. For that reason, I chose 8x42s with 19.6mm of **eye relief,** a measurement that tells you how far the binoculars can be held away from the eye and still present the full field-of-view image in focus. If you don't wear glasses when using binoculars, the eye relief measurement can be in the range of 9–13mm. If you have thin lenses, you may find 14–15mm more comfortable.

Other factors, like **field of view**—the width of the visible area in your glasses—may be important to you. Or maybe you want a really bright pair of binoculars for birding in low light conditions. Look for a wider **exit pupil**—5mm or more is great for dark forests and evening birding.

Don't worry about knowing all the details; just be ready to clearly articulate to the salesperson how you will be using your binoculars, what conditions are most common, and your preferences on the subject of weight, brightness, and focusing.

Quality

How much quality can I afford? There's no point in going for high magnification if you can't spend the big bucks. Magnification requires high-quality optics and sophisticated focusing features that all add up to an expensive pair of binoculars. Consider looking for a lower magnification lens that gives you other important features.

Lens coating is a key element of good binoculars, so ask your salesperson whether the lenses are coated, fully coated, multicoated, or even fully multicoated. Well-coated lenses are less reflective and produce a sharper, brighter image, so this feature is well worth the extra cost.

Don't evaluate **binocular weight** by itself either. Most of the standard binoculars on the market range between 21 and 28 ounces; this isn't really going to make too big a difference. But the weight distribution of that pound and a half is going to make the difference between a comfortable day of birding and a miserable one. A cheap pair of glasses may feel okay at first, but after a long day, you may be ready to throw them out the window!

Another feature that less-expensive binoculars sometimes sacrifice is easy **focus** mechanisms. Central focus is the key to convenience—you don't want to have to adjust each eye individually. Less-expensive binoculars do have central focus. The difference is in how quickly and smoothly the central wheel brings you into sharp focus. Play with this feature in the store and compare models until you find the one that suits you best.

Beyond these features are the bells and whistles. Rubber armor, water- and fog-proof housing, and close-focus capabilities all add cost, but may be worth the convenience. Look for the little details, too, like the quality of the eye cup and lens caps.

All that's left to do is marry your must-haves to your wallet! While it is true that you get what you pay for, not all price points are created equal. There are many good birding binoculars for under $100. The ones I chose were in the mid-range, about $300. But I have my eye on a pair of $1,500 beauties. If I ever get that trip to Africa, those babies are coming along with me!

Fit

How do the binoculars feel in my hand? When it comes right down to it, how comfortable they are to use is what it's all about. Are they bright enough? Do your fingers move easily and naturally to the focus wheel? Is the balance good and the size pleasing? Don't just go with someone else's recommendation; take the time to try out different models. A friend who works the optics counter at the L.L. Bean store in Freeport, Maine, once told me that people rarely choose the binoculars they thought they were going to buy when they arrived at the counter. Once different models are placed in your hand, you quickly gravitate toward the better fit. So plan some time for your binoculars shopping trip; your choice will last you many years, and the right pair will become a trusted friend in the field.

Spotting Scopes

For magnification stronger than 8x or 10x, a spotting scope is the accessory of choice for birders in the field. Spotting scopes are ideal for conditions where distance is an issue. They come in a very wide range of magnifications, from 12x all the way to 75x. Prices vary, too, with an entry-level scope available for a little over $200, and a deluxe version for over $2,000.

The basics of binocular optics more or less apply to spotting scopes. Magnification and lens diameter follow the same conventions, but scope construction differs in a few fundamental ways.

Scope construction can be straight or angled. A **straight scope** is exactly what it sounds like—a straight barrel with an eyepiece at one end. Straight scopes are easy to use when you are seated, or the birds you are viewing are on the ground or at the level of the horizon. The eyepiece is easier to protect in bad conditions. An **angled scope** has an eyepiece set at a 45 degree angle to the barrel. This construction is easier to use if you are viewing high in the sky.

The **eyepiece** is usually sold as a separate accessory to the scope. Eyepieces are available in different magnifications. Some have zoom or wide-angle capabilities, or long eye relief for eyeglass wearers. With such a lot of magnification, look for a focus mechanism that locks into fine focus with a minimum of fuss.

Spotting scopes require the use of a **tripod.** Tripods vary in construction, weight, and height. Your scope needs a strong and steady base to be really effective, so don't skimp on your choice of tripod. Good tripods often require you to make two choices: First, decide on your **tripod base,** which consists of the legs and the center column. These are generally made of aluminum or carbon fiber. The base needs to stand up to adverse conditions and be easy to set up and take down quickly. Strong, stable legs are a must for getting the most out of your scope. Consider, too, how often you will be carrying the scope. Carbon fiber will be lighter than aluminum, but chances are that a good one will be more expensive.

Your next choice is the **tripod head.** For scoping, the head only needs to turn in two directions, up and down or right and left. So don't choose a roll head unless you are going to use your scope for photography. Most tripod heads come with a screw-on attachment. Consider adding a quick-release plate that lets you snap your scope into place. This is an advantage both for speed of assembly and for wear and tear on both the tripod head and the scope's screw mechanism.

Digiscoping is a popular way to shoot birds at long distances. Use a **spotting scope adapter** to attach your camera to a standard spotting scope for photographing birds that may have just appeared as a speck with your regular lens. The quality of digiscoped pictures varies, but some are quite good. Even if your shot is not a prizewinner, this technique is great for capturing and examining birds that you couldn't have gotten a good look at otherwise.

Electronic Guides

These days I also have an "electronic" guide—my iPhone. It is a guide in the most literal sense. Not only does it track my own location and direct me to my next destination with its GPS function, but it also gives me weather updates, let's me text sightings to other birders, and has a handy field guide with birdsong recordings. I use iBird as my field guide. I like the "similar" section that compares possible birds, and I occasionally use the song function to check what I am hearing, or to lure birds a little closer. You can record and share videos, pictures, and lists with other birders, if you like. You can also tweet your sightings or post them to a national tracker. I prefer the iPhone because it lets me do several things at once, but the BlackBerry and other new phones also offer a range of functions and applications. Do a

BELOW: The flat horizon of the beach is the perfect place to use a straight scope. Watching sanderlings run up and down with the tide makes for great entertainment.

little research into the phone styles that your provider offers to get the accessories you prefer.

Top 25 U.S. Birding Locations

Have you exhausted your local haunts, and find that you're looking for adventure? There are many traditional lists of top birding spots, and some of them are included on the list below. But this list represents our twenty-five favorites—a few that we have visited specifically for the birding, some because we discovered them near areas we often travel to, and others simply because of amazing moments we experienced there. It's exciting to develop your own list. Look both near and far for your birding locations—the best ones are the ones you can visit again and again!

1) **Audubon Bird Sanctuary, Dauphin Island, Alabama**
   This sanctuary's 164 acres are uniquely situated to capture some of spring's most spectacular migrations. It is the launch point and landfall for a number of species that migrate back and forth across the Gulf of Mexico. It is fascinating to visit the area since Hurricane Katrina, to watch the adaptation of the habitat and its inhabitants.

2) **Patagonia-Sonoita Creek Preserve, Patagonia, Arizona**
   The best months to visit this rich floodplain valley along the Sonoita River are March through September. One of the top birding spots in the Southwestern United States, more than 300 species live, nest, or migrate through this area.

3) **Point Reyes National Seashore, California**
   Birding at its best—over 45 percent of all North American birds eventually show up at Point Reyes, making it one of the most spectacular birding locations in the world.

4) **J. N. (Ding) Darling National Wildlife Refuge, Sanibel, Florida**
   For us, winter isn't complete without a visit to Sanibel and Ding Darling, with its 6,400 acres of mangrove forest, sea-grass beds, and marshland. The refuge is home to 220 different bird species. Rent a bike and ride through the loop to see herons, egrets, and roseate spoonbills—and make sure to watch for alligators.

5) The Everglades, Florida

A unique environment of saw-grass plain, on average just 6 inches deep, is the ideal habitat for wading birds—grebes, pelicans, wood stork, herons—as well as a wealth of other birds.

6) Jekyll Island, Georgia

A popular resting spot along the Atlantic Flyway, this is a great location to catch the spring and fall migrations, as well as for spotting the local shorebird residents.

7) Snake River Birds of Prey National Conservation Area, Snake River, Idaho

Situated along 80 miles of the Snake River, this area boasts the largest population of nesting raptors in the entire country. Over 800 nesting pairs of falcons, eagles, hawks, and owls nest and raise their young in this unique 485,000-acre area.

8) Chautauqua National Wildlife Refuge, Illinois

This 6,200-acre preserve is located in the middle of the Mississippi Flyway, and during the winter you can find over 450,000 waterfowl and 10,000 shorebirds congregating here.

9) Cimarron National Grassland, Kansas

We find this wild prairie habitat a refreshing change from the forests and seashores where we often go birding. Look for roadrunners, bushtits, curve-billed thrashers, and western tanagers, as well as Cassin's sparrows, lark buntings, scaled quail, and lesser prairie chickens.

10) Grand Isle, Louisiana

This barrier island may have more birds per capita than anywhere else in the U.S. during spring migration—over 180 species have been noted. Grand Isle is home to a spring Migratory Bird Festival every April, and may have given me the greatest weekend of birding I have ever had in my life.

11) Baxter State Park, Maine

This state park at the base of Mount Katahdin boasts 204,000 acres of remote country, perfect for birding and photography. We spent several memorable days here, watching nesting loons on Katahdin Lake. Unfortunately, it is also home to large and aggressive mosquitoes and blackflies.

12) Plum Island, Newburyport, Massachusetts
An amazing area of salt water and marshlands just north of Boston, Plum Island is great in spring for the warblers, and great in winter for the owls, hawks, and shorebirds.

13) Boundary Waters Canoe Area, Minnesota
Imagine a million acres of northern forest paradise free of logging, mining, and motorized vehicles. More than 1,500 miles of canoe trails provide intimate access to loons, mergansers, eagles, and forest songbirds.

14) Noxubee National Wildlife Refuge, Mississippi
This refuge has 48,000 acres of lakes, bottomlands, and pine woods, making it a rich feeding and nesting area for a number of birds, including the endangered red-cockaded woodpecker.

15) Medicine Lake National Wildlife Refuge, Montana
Medicine Lake features one of the largest American white pelican colonies in the United States, with 3,000 to 5,000 nests per year. Centrally located within the breeding range of the majority of prairie songbirds, it's also a popular area for migrating and breeding waterfowl.

16) Red Rock Canyon National Conservation Area, Nevada
Deep sandstone canyons and (relatively) cool conditions make this area a haven of nature near Las Vegas, and a welcome respite from the bright lights of the city. Find over 100 species here, from the tiny cactus wren to the roadrunner, along with a variety of birds of prey.

17) Cape May, New Jersey
One of the top-ten birding spots on everyone's list, Cape May is a unique crossroads for migrating birds, and the best spot on the entire East Coast for hawks that make their way to Cape May every fall. This is a great year-round spot for serious birders.

18) Bosque del Apache National Wildlife Refuge, New Mexico
This is a great winter destination for birders, as tens of thousands of birds, including many kinds of geese, ducks, and sandhill cranes congregate here for winter feeding. Peak visitation in the park occurs in winter, but spring, summer, and fall hold their own delights, with an abundance of warblers, flycatchers, and nesting shorebirds.

19) Central Park, New York City

I love an early-morning walk in Central Park, where warblers, tanagers, robins, and grosbeaks may be glimpsed, along with a variety of waterbirds, and, of course, the ubiquitous rock pigeon. In fact, more than 250 species can be expected to make their way through the park in the course of a year, making it well worth taking an extra day for bird-watching on your next trip to the city.

20) Long Lake National Wildlife Refuge, North Dakota

I'll have to admit, this refuge has never been a destination spot for us, but it does make an interesting stop-off on the way to other western locations. The habitat here is referred to as the Prairie Pothole Region, and contains a mix of marshland and native grasslands. Located as it is in the Central Flyway, it is a great location to see migrating birds of all types, particularly ducks and other waterbirds in the marshes and along the shores of Long Lake.

21) Upper Klamath National Wildlife Refuge, Oregon

Featuring 15,000 acres of freshwater marsh and open water, this area is best explored by canoe for paramount bird viewing. The refuge contains a nesting area for all sorts of waterfowl and colonial nesting birds, including the American white pelican and several heron species, as well as bald eagle and osprey.

22) Hawk Mountain Sanctuary, Pennsylvania

This is one of the best hawk and falcon watching sites in North America, and well worth a trip in October, when the viewing is at its peak.

23) Santa Ana National Wildlife Refuge, Texas

The granddaddy of birding locations, this area of subtropical forest along the Mexican border is home to almost 400 different species of birds. Visit in late January or early February to jump-start your life list.

24) Chincoteague National Wildlife Refuge, Virginia

Come for the ponies, stay for the birds! This area is high on our list, partly due to its proximity to our home. The barrier island is a staging area for many species migrating to South America in the fall, and thousands of waterfowl spend the winter here.

25) Horicon Marsh, Wisconsin

Horicon Marsh is about an hour and a half from the University of Wisconsin and a three-hour drive from Chicago, making it a great weekend destination for hiking, photography, and bird-watching. Horicon is the largest freshwater cattail marsh in North America, and an important resting area for migrating birds. Roger Tory Peterson put it on his list of top American birding locations.

ABOVE: A great egret lands on the water at Chincoteague National Wildlife Refuge.

66 Use the talents you possess—for the woods would be a very silent
place if no birds sang except for the best. **99**

—HENRY VAN DYKE

$$\text{(10)}$$

Five Ways to Make Our
Planet Bird-Friendly

ANYONE WHO takes the time to feed birds or goes out on walks eager to catch a glimpse of a migrating visitor understands how precious and integral the bird population is to the health of the planet. Like the proverbial canary in the coal mine, birds are indicators of environmental conditions—our own personal warning signal when things are not right, whether from farming practices, overdevelopment, or toxic chemicals. We can thank birds for sounding the alarm on DDT, a pesticide that killed off millions of songbirds in the 1960s before scientists noticed it was also getting into the bloodstreams of babies through their mother's milk.

Human encroachment takes away millions of acres of bird habitat every year. That's not just bad for the birds that lose their homes; it's also bad for humans who lose their natural buffers for draining and retaining water, trees that produce oxygen, and clean, flowing water. It is essential to take action to preserve wild areas and to practice sound environmental habits. Certainly it helps the birds, but it's also crucial for us as a population.

Even a series of small changes can contribute to the health and well-being of the planet. These five small steps can have a significant impact on the health of the bird population, both in your backyard and around the world.

Shop Smart

You know that adorable ruby-throated hummingbird sipping on nectar on your back porch right now, as you eat your breakfast? It is possible that he could have recently inspected the coffee beans, bananas, and chocolate on your table. Many species like this one have traveled from their winter grounds in Central or South America. Unfortunately, a number of our favorite foods are grown in the tropics, often in conditions that destroy birds and their habitat. Deforestation, heavy pesticide use, and pollution have taken a heavy toll on the populations that we count on seeing every spring.

Buy organic fruits and vegetables whenever you can. Chemical fertilizers and pesticides aren't good for birds, they aren't good for the environment, and they aren't good for humans! If you don't do another thing to help birds, try to change your buying habits in these simple ways:

- **Bananas.** Go certified. Bananas that have been grown using safe and sane farming practices carry the RAINFOREST ALLIANCE CERTIFIED label. Started in 1991, 15 percent of all bananas are now grown this way. So choose bananas that carry their certification. A few pennies extra will go a long way toward make sure your spring feeder is full of healthy, happy birds just in from the Gulf of Mexico!

- **Coffee.** How many times have I stood looking out at the feeder with a cup of coffee in my hand? Birds and coffee have been part of my morning routine for as long as I can remember. And yet the difference in the way the beans in your cup have been grown can make a huge difference to those warblers singing in your trees right now! Sunny plantations produce high yields, but often require removing trees and using more pesticides and fertilizer. Shade-grown coffee actually creates a great environment for birds, protects soil from erosion, and makes a better-tasting cup of coffee. So buy shade-grown or "bird-friendly" coffee. Coffee that carries a FAIR TRADE CERTIFIED label has been grown on a plantation committed to sound environmental practices, as well as humane treatment of their farm workers. Americans drink 20 percent of all the coffee produced in the world, so imagine the impact we could have by making this simple change!

- **Chocolate.** Chocolate is a smaller industry than coffee, but growing cacao beans produces many of the same difficulties for rainforest residents. Buy organic, or look for the FAIR TRADE or RAINFOREST ALLIANCE certification.

- **Fish and seafood.** Poor fishing practices kill millions of fish every year, and fish farms and heavy metal dredges pollute watersheds. Ask your purveyor about the fish you buy and where it came from. For example, wild Alaskan salmon is preferable to farm-raised or overfished Atlantic salmon. Look for certification from the Marine Stewardship Council, and support grocery chains like Wegmans, which sells MSC-certified sea bass, wild-caught salmon, halibut, Pacific cod, and western Australian lobster tails.

This black-and-white warbler travels widely, often wintering as far south as Venezuela before coming back to breed in North America.

Support Sustainable Forests

It just makes sense that one of the most important ways you can help birds is to support companies that work to keep their habitats safe. That's why it is so important to know where your wood and paper supplies come from, and to buy only from companies that carry certification from the Forest Stewardship Council (FSC). Millions of acres of critical forest habitat have been lost to irresponsible forestry practices, such as clear-cutting and chemical pesticide management. But more companies are beginning to understand that it is in the best interest of their forests, as well as native wildlife, to manage for sustainable growth and biological diversity. Here are some ways you can help:

- **Lumber.** Buy certified lumber from sustainably managed forests. If your lumberyard or home improvement company doesn't carry sustainable lumber, locate and support one that does, and take a few minutes to write to express your desire for sustainable lumber. Particularly avoid tropical hardwoods and old-growth species unless they are certified.

- **Furniture.** Ask where the wood in your furniture came from. Look for finished products certified by the FSC. They certify forest managers as well as production facilities to make sure they meet both social and environmental standards. There are other certification programs, too, like the Rainforest Alliance's SMART WOOD label. Just make sure there is real substance rather than just marketing hype behind the label that you see.

- **Recycled.** Look for reclaimed or recycled wood. Not only does it generally have a lot more character than new wood, but timbers from old barns or buildings come in some amazing widths.

- **Non-wood substitutes.** Skip the wood altogether and use non-wood substitutes. Brands like Trex or SmartDeck, Plastic Lumber, and Polywood are made from recycled plastics. Not only are they good for the forest, but they are also extremely durable and easy to care for. (Just don't use these woods to build nesting boxes! They aren't breathable enough. Stick with real wood for your birdhouse projects.)

- **Recycled fibers.** Is there anything crazier than using virgin timber for products that we use for five seconds and then throw away? Much of the timber for these products comes from Canada's boreal forests, an area that is the breeding ground for up to three billion birds every year. It

ABOVE: This purple martin thanks you for his birdhouse!

is so easy to send a message to these manufacturers. Choose only toilet tissue, paper towels, facial tissues, and other paper products that have high recycled fiber content, preferably without chemical bleaching treatment. The notion that these products may be scratchy or uncomfortable is wrong. If you don't like one brand, try another one. The latest recycled tissues are better than ever. And let the big manufacturers know that you want products that don't destroy the forest.

Certify Your Own Habitat

Do you already practice great yard and woodlot management? Do you grow local plants and practice sustainable gardening? Do you provide bird shelter, cover, and water sources? If so, consider getting your own certification that tells birds as well as your neighbors that your property is a friendly place. The National Wildlife Federation has a program that helps you certify your own land. NWF's Wildlife Habitat Program certifies that you have provided "the essential elements for healthy and sustainable wildlife habitats and have earned the distinction of being part of National Wildlife Federation's Certified Wildlife Habitat program." To qualify for the program, follow the criteria set out on the NWF Web site (visit www.nwf.org for more information).

Even if you don't feel the need to get certified, the practices they outline will make your own habitat good for yourself, your family, and your birds.

Provide food. Plant native shrubs and fruit-bearing plants, and hang feeders for birds and other wildlife. NWF's certification suggests three of the following types of plants or supplemental feeders:

- Seeds from a plant
- Berries
- Nectar
- Foliage/twigs
- Nuts
- Fruits
- Sap

- Pollen
- Suet
- Bird feeder
- Squirrel feeder
- Hummingbird feeder
- Butterfly feeder

Provide water. Birds and wildlife need water, and natural sources are often lacking. Your habitat should have at least one of the following sources to provide clean water for wildlife, for drinking and bathing:

- Birdbath
- Lake
- Stream
- Seasonal pool
- Ocean

- Water garden / pond
- River
- Butterfly puddling area
- Rain garden
- Spring

Provide cover. Birds and other wildlife need shelter from the weather and predators. Your habitat should have at least two areas available for ready cover. Here are some possibilities:

- Wooded area
- Bramble patch
- Ground cover
- Rock pile or wall
- Cave
- Roosting box

- Dense shrubs or thicket
- Evergreens
- Brush or log pile
- Burrow
- Meadow or prairie
- Water garden or pond

Provide nesting areas. Your property should be hospitable to mating birds, and contain at least two areas that support them in raising their young.

- Mature trees
- Meadow or prairie
- Nesting box
- Wetlands
- Cave

- Host plants for caterpillars
- Dead trees or snags
- Dense shrubs or a thicket
- Water garden or pond
- Burrow

Practice safe and sustainable gardening. Here are some options to consider:

- Practice soil and water conservation.
- Control exotic species and reduce lawn size.
- Use chemical-free and organic methods for fertilization and pest control.

Conserve Resources

Everything we do impacts the other living creatures around us. Humans are not the only ones competing for food, water, fresh air, and clean water. Americans are massive consumers of resources. If each of us practiced just a few small steps, it would make a big impact on the environment and the well-being of birds and wildlife.

BELOW: This female rose-breasted grosbeak enjoys a moment at the feeder.

Conserve water. Americans use an astounding 100 gallons of water *every* day. Compared to the rest of the world, whose average is only 22 gallons, it is a shameful statistic. Given that we only need about two quarts a day to actually stay alive; the likelihood that you can find ways to trim back consumption seems high.

- Reduce water usage by watering lawns less.
- Only run full loads in washers and other appliances.
- Install low-flow toilets and showers.
- Be respectful of public sewage systems by refraining from flushing or disposing of products and chemicals that can clog the system and cause contamination in lakes and rivers.

Conserve electricity. Coal-fired plants, hydroelectric and nuclear plants, even wind energy—virtually every single method that humans have for producing electricity is either harmful to the environment or kills birds outright. No one is going to suggest that humans should go without electricity, but you can contribute tremendously to the well-being of birds by reducing the amount of power you need to run your home. There are so many simple things you can do:

- Insulate your water heater.
- Turn lights off when you leave the room.
- Add power strips to all electronic devices and turn them off when not in use. Computers and large-screen televisions waste an enormous amount of power, even when they are turned off.
- Unplug chargers and other household appliances when not in use. Everything that is plugged in uses some power. What a waste!
- Set your thermostat to a reasonable level and leave it there, or replace it with an electronic version that adjusts temperatures automatically throughout the day.

Conserve gas and use alternative forms of transportation. Gas guzzling does more than just hit your pocketbook; it also dumps tons of emissions into the atmosphere, polluting the air and contributing to climate change. Here are some tips:

- Buy a fuel-efficient car.

- Maintain your car properly. A clean air filter, properly inflated tires, and a tuned-up engine all make your car more efficient and keep you from burning your dollars as you head down the road.
- Slow down: Traveling at 55 mph is 15 percent more efficient than 65 mph.
- Carpool.
- Take alternative transportation.
- Use people power! Walking or biking instead of getting into traffic can be a stress reliever—and great exercise.

Recycle your equipment. Had your eye on that new camera or spotting scope, but hate to waste the perfectly good one you already have? No problem. With these options below, you can buy that new one and help out other birders at the same time.

- Donate your old binoculars, camera equipment, tripod, or spotting scope to the Birders' Exchange, an organization that helps to put equipment into the hands of people who need them. They also accept laptops, backpacks, and gently used field guides. Check them out at www.aba.org.
- Optics for the Tropics also collects binoculars to send to ornithologists working in the Caribbean and Latin America (www.opticsforthetropics.org).

ABOVE: Is this northern mockingbird singing for his supper?

Get Involved

If you are an active birding enthusiast, there are lots of ways to get involved. If you're sociable, there are ways to do projects with other birders. If you prefer solitary activities, there are projects for you, too!

Count yourself in. There are many ways to contribute to annual bird censuses. It's fun, and a helpful way to assist ornithologists and other scientists as they track the health of bird populations across the country.

* Get together as a group to do the Annual Christmas Bird Count. For over a hundred years, volunteers have gotten together for three weeks, from mid-December to early January, to participate in the National Audubon Society's Annual Count. Go to www.audubon.org to learn how to get involved.
* Participate at your own window in the Great Backyard Bird Count. The GBBC is a four-day event, created specifically to get real-time feedback on the population and distribution of birds all over the country. A joint project of the Cornell Lab of Ornithology and National Audubon, this event takes place every year in mid-February. To sign up, go to www.birdsource.org/gbbc.

- Join NestWatch—a continent-wide nest-monitoring database of the Cornell Lab of Ornithology, funded by the National Science Foundation and developed in collaboration with the Smithsonian Migratory Bird Center. NestWatch teaches people about bird-breeding biology and engages them in collecting and submitting nest records. For more information, go to www.nestwatch.org.

Help out at a nature center or wildlife refuge. Every school, nature center, and park in the country is in constant need of able-bodied volunteers of every kind. No matter what your interest or skill level, there is a job waiting for you!

- Volunteer to speak to school groups or to lead children's birding walks.
- Join a day of trail clearing or nesting-box maintenance.
- Work in the park's gift shop or in the office.
- Volunteer to present educational programs to service groups, clubs, and retirement communities. A friend of mine offered to install and maintain a feeder at her local nursing home. She set it up outside the dining room windows, much to the delight of the residents.

Donate to groups doing good things. There are a lot of ways to help, and cash donations are always welcome. Think locally as well as nationally when considering who might be on your donation list. Although we are members of some national organizations, we get the most satisfaction from the small donations we give to our local parks and preserves.

- Buy the Migratory Bird Hunting and Conservation Stamp, also known as the duck stamp. Fees for this stamp go toward acquiring land for national wildlife refuge lands, and give you free admission into any refuges that charge for entry.
- Support our national parks by donating to your favorite one, or join a "Friends of the Park" group.
- Remember your local parks, preserves, and civic groups.
- Make conservation organizations a regular part of your charitable giving.

Resources

Field Guides

Brinkley, Edward S., and Craig Tufts. *National Wildlife Federation Field Guide to Birds of North America.* New York: Sterling Publishing, 2007.

Bull, John L. *National Audubon Society Field Guide to North American Birds: Eastern Region.* New York: Knopf, 1994.

Cornell Laboratory of Ornithology, and Roger Tory Peterson. *A Field Guide to Bird Songs: Eastern and Central North America* (audiobook). New York: Houghton Mifflin Harcourt, 1990.

Dunn, Jon L., and Jonathan Alderfer. *National Geographic Field Guide to the Birds of North America, Fifth Edition.* Washington, D.C.: National Geographic Society, 2006.

Floyd, Ted. *Smithsonian Field Guide to the Birds of North America.* New York: Harper Paperbacks, 2008.

Kaufman, Kenn. *Kaufman Field Guide to Birds of North America.* Boston: Houghton Mifflin Harcourt, 2005.

Peterson, Roger Tory. *Peterson Field Guide to Birds of North America (Peterson Field Guide Series), 100th Anniversary Edition.* Boston: Houghton Mifflin Co, 2008.

Sibley, David Allen. *The Sibley Guide to Birds*. New York: Knopf, 2000.

Stokes, Donald, and Lillian Stokes. *Stokes Field Guide to Birds: Eastern Region*. New York: Little, Brown and Company, 1996.

Stokes, Donald, and Lillian Stokes. *Stokes Field Guide to Birds: Western Region*. New York: Little, Brown and Company, 1994.

Thompson, III, Bill, and Julie Zickefoose. *The Young Birder's Guide to Birds of Eastern North America (Peterson Field Guides)*. Boston: Houghton Mifflin Harcourt, 2008.

Udvardy, Miklos, and John Farrand. *National Audubon Society Field Guide to North American Birds: Western Region*. New York: Knopf, 1994.

Backyard Bird Books

Burton, Robert, and Steven Kress. *The Audubon Backyard Birdwatcher: Bird-feeders and Bird Gardens*. Chicago: Thunder Bay Press, 2002.

Heintzelam, Donald. *The Complete Backyard Birdwatcher's Home Companion*. Camden, ME: International Marine Press, 2000.

Roth, Sally. *The Backyard Bird Feeder's Bible: The A-to-Z Guide to Feeders, Seed Mixes, Projects, and Treats*. New York: Rodale Press, 2003.

Schneck, Marcus H. *The All-Season Backyard Birdwatcher: Feeding and Landscaping Techniques Guaranteed to Attract the Birds You Want Year Round*. Dallas: Quarry Press, 2005.

Sill, Cathryn, and John Sill. *About Birds: A Guide for Children*. Atlanta, GA: Peachtree Publishers, 1997.

Recommended Reading

Carson, Rachel. *Silent Spring*. Boston: Houghton Mifflin, 1962.

Louv, Richard. *Last Child in the Woods: Saving Our Children from Nature-Deficit Disorder*. Chapel Hill, NC: Algonquin Books, 2008.

Rogers, Elizabeth, and Thomas M. Kostigen. *The Green Book: The Everyday Guide to Saving the Planet One Simple Step at a Time*. New York: Three Rivers Press, 2007.

Peterson, Roger Tory. *All Things Reconsidered: My Birding Adventures*. Boston: Houghton Mifflin Harcourt, reprint 2007.

Urquhart, Thomas. *For the Beauty of the Earth: Birding, Opera and Other Journeys*. New York: Counterpoint, 2006.

Weidensaul, Scott. *Of a Feather: A Brief History of American Birding.* Boston: Houghton Mifflin Harcourt, 2007.

Weiner, Jonathan. *The Beak of the Finch.* New York: Vintage Books, 1995.

Magazines for Birders

Birder's World	*Bird Watcher's Digest*
P.O. Box 1612	P.O. Box 110
Waukesha, Wisconsin 53187	Marietta, Ohio 45750
www.birdersworld.com	www.birdwatchersdigest.com
Birds and Blooms	*WildBird Magazine*
5400 S. 60th Street	P.O. Box 6050
Greendale, Wisconsin 53129	Mission Viejo, California 92690
www.birdsandblooms.com	www.wildbirdmagazine.com

Online Resources

American Bird Conservancy	National Audubon Society
www.abcbirds.org	www.audubon.com
American Birding Association	National Wildlife Federation
www.aba.org	www.nwf.org
Boreal Songbird Initiative	National Wildlife Refuge Association
www.borealbirds.org	www.refugenet.org
Cornell Lab of Ornithology	Rainforest Alliance
www.allaboutbirds.org	www.rainforest-alliance.org
Forest Stewardship Council	WhatBird: The Ultimate Bird Guide
www.fscus.org	www.whatbird.com

Birding Organizations and Clubs

Alabama

Alabama Ornithological Society

Founded in 1952. Web site has info about birds and birding in Alabama, checklist of Alabama birds, bird-count data, membership info, schedule of meetings, and Rare Bird Alert reports for Alabama and northwest Florida.

Alaska

Mat-Su Birders

A local wild bird club in the beautiful Matanuska-Susitna Valleys in Alaska, Mat-Su Birders formed in January 1999. Members are folks from all walks of life, all age groups, all levels of bird knowledge and birding experience from beginner to expert, old-timers and newcomers to the area. Everyone is welcome!

Arizona

Desert Rivers Audubon Society

Covers the eastern portion of Maricopa County.

Maricopa Audubon Society (MAS)

Phoenix metropolitan-area chapter, with over 3,000 members. Web site has info about the chapter, programs and meetings, online newsletter, links. This club covers the central portion of Maricopa County.

Sonoran Audubon Society

Covers the western portion of Maricopa County.

Tucson Audubon Society (TAS)

Working in southern Arizona for more than fifty years to promote conservation and environmental education, and to provide environmental recreation.

Arkansas

Audubon Society of Central Arkansas (NAS chapter)

Web site includes info on meetings, citizen science programs, and a map with directions to birding hot spots.

Northwest Arkansas Audubon Society (NAS chapter)

Web site has a growing bird list.

California

Kern Audubon Society

Serves the Southern San Joaquin Valley and the Greater Bakersfield area. Web site has info relevant to the environment, birding, education, conservation, and nature lovers.

Sea & Sage Audubon Society

Serves northern, central, and a portion of southern Orange County. Web site has info about field trips, conservation activities, programs, birding classes for adults, nature programs for children, volunteer opportunities, and a library of nature slides.

Tehachapi Mountains Birding Club

Serves birders in the southern Sierras community of Tehachapi. The Web site's primary focus is educational, and includes local bird species articles, reports on several club research projects (including the annual Turkey Vulture migration census), birding book reviews, bird photos, and schedule of meetings and field trips.

Colorado

Colorado Field Ornithologists

The official state organization is dedicated to the field study, conservation, and enjoyment of Colorado birds. Web site contains info about the organization as well as Colorado birding.

Florida

Audubon Society of the Everglades

Incorporated in 1966 and serves communities from Jupiter and Tequesta south to Boca Raton.

Bay County Audubon Society

Founded in 1962 by citizens dedicated to the preservation of natural areas and wildlife habitat. Serves Panama City and surrounding areas in Bay, Gulf, Calhoun, Washington, Holmes, and Jackson counties.

Georgia

Georgia Ornithological Society

Web site has membership info, Georgia and S. Georgia / N. Florida Rare Bird Alert transcripts, trip reports, species accounts, Christmas Bird Count and North American Migration Count data, and links for neighboring states, plus a photo index.

Illinois

North Central Illinois Ornithological Society

Rockford's Bird Club has been around for sixty years.

Iowa

Iowa Ornithologists' Union (IOU)

Hosts Iowa Birds & Birding Web site, with info on IOU, Iowa Rare Bird Alert , and Iowa Chat, birding locations, hot birds, checklists, Christmas Bird Counts, and useful local contacts.

Kansas

Kansas Ornithological Society (KOS)

Web site contains information about KOS, upcoming events, and back issues of the quarterly newsletter, *The Horned Lark*. The site also provides current Kansas checklist; individual county checklists for all 105 Kansas counties; links to the Kansas Bird Discussion list archives; and a list of special KOS projects. KOS was formed in 1949 and serves nearly 400 members.

Kentucky

Kentucky Ornithological Society

Web site includes the Kentucky Rare Bird Alert, a checklist of Kentucky birds, plus info about *The Kentucky Warbler* (the official publication of KOS), membership, meetings, field trips, bird counts, and bird conservation, plus links to numerous other bird-related sites.

Louisiana

Baton Rouge Audubon Society
Crescent Bird Club

Formed in 1966 to serve the New Orleans area, which, being at the southern end of the Mississippi Flyway and on the Gulf of Mexico, gets a wide variety of species. Club sponsors free birding trips for members to birding hot spots.

Gulf Coast Bird Club (Lake Charles)

Louisiana Ornithological Society

Formed in 1947 to gather and disseminate accurate information about the birdlife of Louisiana and the western hemisphere. Web site includes info about the society's history, officers, membership, meetings, and trips; state birding organizations; a state checklist; the Louisiana Bird Records Committee; LOS News, Louisiana Birdline, and local contacts.

Northeast Louisiana Bird Club (Monroe)

Northshore Bird Club (Slidell/Covington)

Orleans Audubon Society (New Orleans)

Shreveport Society for Nature Study

Subgroup of the Bird Study Group, Shreveport's organization of bird-watchers. Offers field trips, bird discussions, a Bird Report database that tracks bird sightings, and other programs of interest to people with an interest in birds.

Maryland

Howard County Bird Club

A chapter of the Maryland Ornithological Society. The club recently published *Birding Howard County, Maryland*, a 134-page spiral-bound guide to finding birds in the county. Web page provides local birding updates and county bird list.

Maryland Bluebird Society (MBS)

MBS was recently incorporated as a nonprofit promoting the conservation of the eastern bluebird through education, the establishment of trails, and support of research.

Maryland Ornithological Society

Founded in 1945, the Maryland Ornithological Society promotes the study and enjoyment of birds. Birding activities, publications, and programs are available to members through the state organization or a local chapter. The Web site is a complete resource on birding in Maryland.

Massachusetts

Brookline Bird Club

Founded in 1913, the club now has more than 1,300 members nationwide. Web page is useful for anyone interested in birding in New England, particularly Massachusetts.

Cape Cod Bird Club

A 500-member club founded in 1971 that offers monthly meetings and weekly field trips. The Web site is useful to anyone interested in birding on Cape Cod.

Eastern Massachusetts Hawk Watch

An all-volunteer, member-based organization whose mission is to promote the study and conservation of hawks locally and on a continental scale by monitoring migration in Massachusetts. The Web site contains information on all aspects of hawk watching.

The Essex County Ornithological Club (ECOC)

Established in 1916 to promote ornithological study throughout Essex County, Massachusetts. The club maintains a long tradition of an annual May bird census by canoe, kayak, and on foot along the Ipswich River. The ECOC periodically revises the official checklist of the Birds of Essex County. Web site includes meeting/speaker schedule and the Field List of the Birds of Essex County, Massachusetts.

Hampshire Bird Club

Based in Amherst. Web site includes info about membership, meetings, field trips, Hampshire lending library book list, and a page of photos.

Hoffmann Bird Club

Founded in 1940, the Hoffmann Bird Club has approximately 125 members and offers monthly meetings and field trips within Berkshire County and to surrounding states.

Massachusetts Audubon

The largest conservation organization in New England, concentrating its efforts on protecting the nature of Massachusetts for people and wildlife. The organization operates thirty-six wildlife sanctuaries across the state that are open to the public and serve as a base for its conservation, education, and advocacy efforts.

Michigan

Erie Shores Birding Association of Monroe

Organized in 1986 for the purpose of promoting the observation and study of birds and other related wildlife, their environment, and their conservation.

Huron Valley Audubon Society of Livingston County

Nonprofit affiliated as an official chapter of the Michigan Audubon Society but completely self-supporting. Goal is education by supporting the process of lifelong learning, with a program of informative meetings, nature-oriented field trips, and conservation projects. Web site includes meeting/speaker schedule.

Michigan Audubon

Mission is to connect birds and people for the benefit of both through environmental research, conservation, and education in the state. The society has about 2,000 members statewide, 41 local chapters, 19 sanctuaries, and 4 affiliates. Michigan Audubon also nurtures partnerships with National

Audubon, Detroit Audubon, and the Kalamazoo Nature Center.

Michigan Loon Preservation Association

Mission is to conserve and enhance the common loon population through research, habitat protection and restoration, species protection, and public awareness and involvement. Web site includes info about the common loon and the organization.

Minnesota

Minnesota Ornithologists' Union

Organization of professionals and amateurs interested in birds. Publishes a journal, *The Loon*, and a newsletter, *Minnesota Birding*. Info-rich Web site includes lists and distribution maps of the birds found in Minnesota, a list of sixteen quality birding locations, and an account of recent rare, unusual, or interesting sightings.

Mississippi

Mississippi Coast Audubon Society

A National Audubon Society chapter. Web site includes a calendar of activities, bird list, and other resources.

Missouri

Webster Groves Nature Study Society

Organization of amateur naturalists interested in plants, insects, and birds of the St. Louis area. Web site includes info on where to find a Eurasian tree sparrow in the area, phone numbers for recent rare bird sightings, field trips, checklist, and archive of recent bird reports and articles from *Nature Notes*.

Nebraska

Audubon Nebraska

The state office of National Audubon Society in Nebraska. Web site has info on NAS activities in the state, plus sections on Audubon's two Nebraska nature centers, Audubon Spring Creek Prairie and Lillian Annette Rowe Sanctuary.

Nebraska Ornithologists' Union (NOU)

Dedicated to the study, appreciation, and protection of birds. Web site gives info about NOU (membership, officers, publications, records committee), plus upcoming birding events, RBAs, and favorite birding areas within the state.

Wachiska Audubon Society

A local chapter of National Audubon Society serving seventeen counties in Southeast Nebraska. Web site features info on field trips, educational meetings, and wildlife habitat conservation.

New Jersey

Gloucester County Nature Club

Local nonprofit focused on educating the public about the natural world around them. Web site includes details on the history, field trips, programs, and special events of this dynamic organization.

Monmouth County Audubon Society

Web site includes news on upcoming programs and events, local news and environmental info, injured bird info, links, and bimonthly newsletter.

New Jersey Audubon Society (NJAS)

Founded in 1897, and one of the oldest independent Audubon societies, NJAS has no connection with the National Audubon Society. Web site includes a birders forum and seasonal site guides for great birding spots in New Jersey.

New York

Brooklyn Bird Club

Comprehensive Web site for New York City area. Lists hot spots in and around Brooklyn and Queens, including Prospect Park, Jamaica Bay Wildlife Refuge, Gateway NRA. Area checklists and maps included. Useful compendium for birders visiting NYC area.

New York State Bluebird Society

A nonprofit dedicated to the conservation of the eastern bluebird (*Sialia sialis*), the state bird. The Web site has links to other bluebird pages.

North Carolina

Piedmont Bird Club

Founded in 1938, the club now has 138 members. Web site has calendar of meetings and field trips, checklist, and information about regional birds, field trip reports, places to bird in Guilford County, North Carolina, photos by members, and links.

Ohio

Kirtland Bird Club

Meets at 7:30 PM on the first Wednesday of each month, September through June, at the Cleveland Museum of Natural History. Web site includes info about meetings, field trips, birds sighted, and a newsletter (in PDF format).

Ohio Ornithological Society

Welcoming backyard birdwatchers and researchers in the field alike, the Ohio Ornithological Society is the only statewide organization specifically devoted to fostering a deeper appreciation of wild birds, fellowship and collabo-

ration in advancing our collective knowledge about them, and our ability to speak with one voice to preserve Ohio's bird habitats.

Ohio Young Birders Club

A chapter of the Black Swamp Bird Observatory, this club was founded in 2006 and serves young birders, ages twelve to nineteen. The club's goals include creating a community for young birders throughout Ohio and beyond; fostering an interest in natural history; and encouraging young people to spend more time outside. The Web site includes information about meetings, field trips, and membership.

Western Cuyahoga Audubon Society

Programs are held on the first Tuesday of the month, September through June, at the Rocky River Nature Center in North Olmsted, Ohio. Field trips are held in the Northern Ohio area. Web site includes programs, field trips, newsletters, animal rehab directory, and volunteer opportunities.

Oklahoma

Audubon Society of Central Oklahoma, Oklahoma City

Bartlesville Audubon Society, Bartlesville

Cleveland County Audubon Society, Norman

Deep Fork Audubon Society,
Shawnee
Grand Lake Audubon Society,
Grove
Indian Nations Audubon Society,
Muskogee
Payne County Audubon Society,
Stillwater
Tulsa Audubon Society, Tulsa
Washita Valley Audubon Society,
Pauls Valley

Oregon

Klamath Basin Audubon Society
The Klamath Basin Audubon Society was founded in 1983 after the Bald Eagle Conference, ongoing since 1979, made local folks see the need for such an organization.

Oregon Field Ornithologists (OFO)
Founded in 1980, OFO publishes *Oregon Birds*, a quarterly journal. Web site has info about OFO, Oregon Breeding Bird Atlas project, Oregon Bird Records Committee, field trips and meetings, state checklist, and links to Oregon Birders Online and other Oregon and youth birding sites.

Pennsylvania

Three Rivers Birding Club
Informative Web site includes info on recent bird sightings, birding hot spots with directions, outings, events calendar, and discussion forums.

South Carolina

Greenville County Bird Club
Serves upstate South Carolina and nearby western North Carolina. Founded in February 2000 to promote birding and to encourage responsible environmental stewardship. Site contains News and Outings sections, as well as trip reports with photographs. Checklists are available for download, and hawk migration counts are available for each year since 1988.

Hilton Pond Center for Piedmont Natural History
Nonprofit based near York. Web site includes photos and info about flora and fauna found in most habitats in the eastern U.S, the Center's long-term bird-banding research projects, and "This Week at Hilton Pond," a pictorial account of natural changes through the seasons.

South Dakota

Sioux Falls Bird Club
An organization founded to promote the enjoyment, study, and conservation of wild birds in the Sioux Falls region.

Tennessee

Tennessee Ornithological Society (TOS)
Founded in 1915, with the first chapter established at Nashville, with Knoxville in 1923. Publishes

quarterly journal *The Migrant* (launched in 1930), plus semi-annual newsletter, the *Tennessee Warbler*. Web site includes info about TOS, birding in Tennessee, upcoming events, Tennessee Breeding Bird Atlas, and Bird Records Committee.

Texas

Houston Audubon

Currently celebrating over forty years of conservation, Houston Audubon promotes the conservation and appreciation of birds and wildlife habitat. Current information on field trips, membership meetings, and other events can be found at www.houstonaudubon.org.

Texas Ornithological Society

TOS was founded in 1953 as a nonprofit organization. The purpose of the society is to promote the discovery and dissemination of knowledge of birds; to encourage specifically the observation, study, and conservation of birds in Texas; to encourage the formation of local birding clubs; and to stimulate cooperation among professional ornithologists. The Web site, www.texasbirds.org, has quarterly bird reports from each of the eight TOS regions and information on upcoming events, including meetings and field trips.

Utah

Utah Ornithological Society (UOS)

The Web site has information about membership, journal subscription, local and state checklists, and birding records in Utah, including access to the records of the Utah Bird Records Committee.

Virginia

Monticello Bird Club

Named after Thomas Jefferson's home on the "little mountain" near the city of Charlottesville.

Northern Virginia Bird Club

The club was founded in 1955 as a chapter of the Virginia Society of Ornithologists. Web site offers schedule of walks (two per week, September through June), newsletter, schedule of meetings, pictures taken by members, and links to other birding sites.

Washington

Washington Ornithological Society

Founded in 1988, "to provide a forum for birders from throughout the state to gather and share information on bird ID, population status and birding sites." Over 500 amateur and professional ornithologist members. Web site includes checklist, photos and links.

Washington State: Rainier Audubon Society (South King County Chapter)

Web site has a lot of info about the society's activities, projects, CBC data, bird sightings, how to subscribe to the Wa-Rainier List-serv, nest boxes, wildbird care, green gardening.

West Virginia

The Brooks Bird Club

An independent, educational nonprofit which promotes the study and enjoyment of birds and other elements of the natural world. Headquartered in Wheeling, West Virginia.

Wisconsin

Bluebird Restoration Association of Wisconsin

Seeks to monitor and increase the population of the eastern bluebird and other native cavity-nesting birds through a coordi-nated statewide nest-box construc-tion and monitoring program.

Chequamegon Bird Club

A club based in Medford, Wisconsin, with over 100 members.

Oshkosh Bird Club

Founded in May 1970 in order to stimulate interest in and promote the preservation of birds. Club members are actively involved each year in field trips, bird counts, and local bird projects.

The Wisconsin Society for Ornithology

Organized in 1939 to encourage the study of Wisconsin birds. The aims have expanded to emphasize all of the many enjoyable aspects of birding, and to support the research and habitat protection necessary to preserve Wisconsin birdlife. Members include back-yard birders, ornithology hobby-ists, as well as professional orni-thologists.

Birding Supplies

There is really no substitute for your local bird-supply store or garden nursery to get the most useful advice about the birds in your area, along with the best native plants. But sometimes online shopping is the most convenient way to locate a specialty item or to research a particular need. These sites are some of my favorites.

Duncraft
Concord, New Hampshire
(888) 879-5095
www.duncraft.com
A family business since 1952, Duncraft offers feeders, nesting boxes, and more.

ebirdseed
(866) 324-7373
www.ebirdseed.com
Located near Fargo, North Dakota, supplying farm-fresh seed and live mealworms from local growers.

Native Plant Nursery Directory
www.plantnative.org
A fantastic resource for information on plants native to your area, as

well as a directory of local resources for native plants and seeds.

Optics4Birding
(877) 674-2473
www.optics4birding.com
Birders who know their stuff when it comes to all things related to birding optics. Shop here for binoculars, scopes, and even night vision gear for after-hours owling.

Plow and Hearth
Madison, Virginia
(800) 494-7544
www.plowhearth.com
Feeders, birdhouses, birdbaths, and decorative accessories for the birding enthusiast.

Wild Birds Unlimited
Carmel, Indiana
(888) 302-2473
www.wbu.com
A chain of birding stores with a great Web site that provides good products and information.

LEFT: Great egret

Index